Understanding Chemical Patents

Understanding Chemical Patents

A Guide for the Inventor
Second Edition

John T. Maynard
Howard M. Peters

ACS Professional Reference Book

American Chemical Society, Washington, DC 1991

Library of Congress Cataloging-in-Publication Data

Maynard, John T., 1919–1977
  Understanding chemical patents: a guide for the inventor
      John T. Maynard, Howard M. Peters. — 2nd edition
          p.   cm.

      Includes bibliographical references and index.

      ISBN 0–8412–1997–4. — ISBN 0–8412–1998–2 (pbk.)

      1. Patents. 2. Chemistry—Patents.
      I. Peters, Howard M., 1940–   . II. Title.

T211.M39   1991
660'.0272—dc20   91–24124
                                                  CIP

The paper used in this publication meets the minimum requirements of American National Standard for Information Sciences—Permanence of Paper for Printed Library Materials, ANSI Z39.48–1984.   ∞

PRINTED IN THE UNITED STATES OF AMERICA

Dedication

This second edition is dedicated to my parents, Elmer and Marie Peters, who made many things possible; to my wife, Sally, who has always provided me with encouragement and support; and to my daughters, Theresa and Elizabeth, who have done their part.

—H. M. P.

About the Authors

JOHN T. MAYNARD was involved in the field of patents during his entire professional life. His work as a research chemist resulted in several publications and 14 U.S. and foreign patents. He joined the Patent Service of E. I. du Pont de Nemours and Company's Elastomer Chemicals Department in 1964 and served as its head from 1970 to 1977. He was also Chairman of the American Chemical Society Committee on Patent Matters and Related Legislation from 1968 to 1973.

Maynard studied chemistry at the California Institute of Technology and received his Ph.D. from Yale University in 1946. He then joined Du Pont, where he remained for 31 years, rising from research chemist to division head. In addition to his work on the Patent Committee, he was chairman of the Delaware Section of the ACS in 1962 and an ACS councilor.

Maynard felt that writing this book (the first edition) was one of the most satisfying experiences of his life. He died suddenly in 1977.

HOWARD M. PETERS has also been involved with chemical research and related matters all his professional life. He performed his first patent research at 21 during a summer research program in 1962 at the Gulf Oil Corporation Research Center near Pittsburgh, PA. In 1966, he joined the Dow Chemical Company of Midland, MI. From 1969 to 1978 he was a research chemist and a project leader at the Stanford Research Institute (now SRI International), in Menlo Park, CA. He is copatentee of seven U.S. patents and is the author of more than 20 chemical publications.

Peters was a staff and patent attorney with the Hexcel Corporation in San Francisco and a patent attorney with Syntex Corporation in Palo Alto, CA. He moved to private practice in patent law with Burns, Doane, Swecker, and Mathis of Alexandria, VA, and San Francisco, CA, and is presently a partner in the patent law firm of Phillips, Moore, Lempio & Finley of San Francisco.

Peters studied chemistry at Geneva College in Beaver Falls, PA (B.S., 1962, magna cum laude), and received his Ph.D. in organic chemistry from Stanford University in 1967 under the direction of Harry S. Mosher. He earned his Juris Doctor in 1978 at the University of Santa Clara, Santa Clara, CA. In addition to his service on the ACS Patent Committee, he served as chairman of the Santa Clara Valley Section of the ACS in 1986, has served as its councilor, and has received its Ottenberg Service Award (1984). He was a cofounder of the ACS Division of Chemistry and the Law in 1979, and is presently serving as its councilor and on its executive committee. Peters received the ACS Division of Chemistry and the Law's Roger D. Middlekauf Service Award in 1991.

Contents

List of Figures and Tables

Preface

Questions about patents most often concern those who are actively working in the chemical profession; consequently, this is a book for practicing chemists and chemical engineers. Much of the information will also be useful to people in other fields who need to understand our patent system. The first edition of this book, written by John T. Maynard alone, was an outgrowth of a short course offered to the technical staff of the Du Pont Experimental Station in 1976. The attendance at that course was the largest of any course ever offered to that audience, and it underscored the widespread interest of chemical practitioners and their need to know more about patents.

This book is not about patent law or patent licensing and management. In this book, Maynard and I try to answer the immediate, practical questions of chemists and engineers about how to read and to understand patents, how to use patents as a source of information, how to recognize that an invention has been made, how to work with attorneys or agents in seeking patent protection for inventions, how to keep adequate notebook records, how to watch for infringement of patents, and so on. Another main purpose of this book is to give the technical person enough familiarity with the special terminology of patents to be able to deal comfortably with patent attorneys, agents, and technical liaison personnel.

Many points about the law and the business aspects of patents have been given only a paragraph or even a single sentence here but are the subject of entire books or journal articles in the literature. For those who want to pursue such matters, the Bibliography suggests a number of books and articles that can be consulted, and the references cited will lead to more.

This second edition of the book relies heavily on the first and updates those many areas that have undergone large changes over the past 12 years. Major changes are still being examined for U.S. and foreign patent law.

A Special Note

To those who have just finished academic training and are starting their professional careers:

You are leaving a world of grades and grading in which a typical scale would rate a score of 70% as passing, 80% as good, 90% as excellent, and 100% as impossible. You are entering a world where the only creditable score, at least in matters of technical accuracy, is 100%. A score of 90% will receive a cool reception, and anything less is unsatisfactory. There is no reason for terror, however, because you are well prepared and you will find that many redundancies are provided in the procedures you will be following. It will be difficult to err very much.

Nowhere in your new activities are error-free results more important than in patent matters. By your efforts, important intellectual property rights are secured. By your acts, important patent rights can also be lost or weakened. Failure to secure sound patent protection for the inventions you will make can result in wasted research and development effort and even substantial liability for your organization if it should infringe the patents of others.

You should take patents seriously, take advantage of your patent information sources, and work closely with your patent advisors and legal representatives when their assistance is appropriate. Much of this book was written with your particular needs in mind. I hope it contributes to smoothing the path in your new career.

Acknowledgments

This volume would be incomplete without acknowledging the helpful assistance of the staff of the Books Department of the American Chemical Society, and my law partners and co-workers in the firm of Phillips, Moore, Lempio & Finley, who provided me with technical support.

A special thanks is due the members of the ACS Joint Board–Council Committee on Patents and Related Matters. This committee gave me the opportunity and time to write and to complete this project. Many committee members read and commented on parts or all of the text.

My former corporate employers, Hexcel and Syntex, and my present clients, including the University of California, Dow Chemical U.S.A., SRI-International, the U.S. Department of Energy, NASA–Ames Research Center, Stanford University, and Osaka Gas Company, are profusely thanked for providing me with many interesting technical and legal situations to continue my growth in patent law.

HOWARD M. PETERS
Phillips, Moore, Lempio & Finley
177 Post Street
San Francisco, CA 94108–4731

June 1991

1

Introduction: The Purpose of Patents

Human beings are competitive creatures. The evolutionary thrust toward survival of the fittest has made us that way, and contemporary people have carried the competitive urge to its highest form. The urge to compete can be seen in all aspects of human activity, from the conflicts of 2-year-olds over possession of a toy through all the physical and intellectual competitions of school years to the ultimate struggles between races and nations for power and possessions. People learned early, however, that the forces of competition must be balanced with an equal measure of cooperation if all of us are not to be destroyed by the process. Organized society resulted, and successful societies have been those that have achieved the most effective balance of competition and cooperation.

Patent systems are among the most straightforward examples of such a successful, cooperative arrangement and show society operating at its best. Patent systems reward the competitive, creative drive with a temporary, limited, exclusive right, in return for the cooperation of an inventor in teaching the rest of society how to use his or her findings for all time thereafter. This philosophy must be kept in mind in order to use and to understand patents effectively. It will be referred to throughout this book.

Early Patent Systems

Patent systems more or less as we know them have been used as a social tool since the Renaissance. Rulers of the Italian city-states recognized that their creative subjects needed encouragement and protection against copying by their competitive fellows. Patents were granted for various terms, the most famous being a 20-year patent granted to Galileo by the Doge of

1997–4/91/0001/$06.00/1

Venice on a mechanism for raising irrigation water to fields. European and English monarchs granted patents to encourage commerce and, unfortunately, to reward favorites of the court. Abuses of the latter sort, in which exclusive rights were granted or sold to individuals to trade in particular commodities or services regardless of their inventive contributions, resulted in adoption of the Statute of Monopolies by the British Parliament in 1624. This law, which required that patents be granted only for genuinely novel contributions to society, is generally regarded as the basis for the present patent system of all countries of the world.

The U.S. Patent System

The American colonies made a practice of granting patents to their inventors, and the framers of the United States Constitution provided the basis for U.S. patent and copyright laws in Article 1, Section 8:

> The Congress shall have the power . . . to promote the progress of science and useful arts, by securing for limited times to authors and inventors the exclusive right to their respective writings and discoveries.

This Constitutional directive was given life by the Patent Act of 1790, which provided for examination of patent applications by a commission consisting of the Secretary of State (Thomas Jefferson), the Secretary of War (Henry Knox), and the Attorney General (Edmund Randolph). The first U.S. patent (Figure 1-1), signed by George Washington, Randolph, and Jefferson, was in fact on a chemical subject. The inventor was Samuel Hopkins of Philadelphia, and his contribution to society was an improved method of making potash from wood ashes for use in soap making. The distinguished commission made a diligent effort to meet their commitment to examine all applications for patentable merit, but within 3 years it became clear that they simply could not give adequate study to every application being submitted. In 1793 a "registration" system was adopted by the United States; patents were issued on every application that met formal requirements, and the burden of sorting out priority of rights among those holding conflicting patents was thrown into the courts.

After all the records from this period of turmoil were destroyed by a fire in the Patent Office in 1836, Congress passed the Patent Act of 1836. This act created the position of Commissioner of Patents and a Patent Office staffed sufficiently to examine applications to determine whether a truly new invention had been made. Present U.S. patents are numbered

The United States.

To all to whom these Presents shall come. Greeting.

Whereas Samuel Hopkins of the city of Philadelphia and State of Pennsylvania hath discovered an Improvement, not known or used before, such Discovery, in the making of Pot ash and Pearl ash by a new Apparatus and Process, that is to say, in the making of Pearl ash 1st by burning the raw Ashes in a Furnace, 2d by dissolving and boiling them when so burnt in Water, 3d by drawing off and settling the ley, and 4th by boiling the ley into Salts which then are the true Pearl ash, and also in the making of Pot ash by fluxing the Pearl ash so made as aforesaid; which Operation of burning the raw Ashes in a Furnace, preparatory to their Dissolution and boiling in Water, is new, leaves little Residuum, and produces a much greater Quantity of Salt: These are therefore in pursuance of the Act, entitled "An Act to promote the Progress of useful Arts", to grant to the said Samuel Hopkins, his Heirs, Administrators and Assigns, for the Term of fourteen Years, the sole and exclusive Right and Liberty of using and vending to others the said Discovery, of burning the raw Ashes previous to their being dissolved and boiled in Water, according to the true Intent and Meaning, of the Act aforesaid. In Testimony whereof I have caused these Letters to be made patent, and the Seal of the United States to be hereunto affixed Given under my Hand at the City of New York this thirty first Day of July in the Year of our Lord one thousand seven hundred & Ninety.

G Washington

City of New York July 31st 1790. –

I do hereby certify that the foregoing Letters patent were delivered to me in pursuance of the Act, entitled "An Act to promote the Progress of useful Arts," that I have examined the same, and find them conformable to the said Act.

Edm: Randolph Attorney General for the United States.

(Endorsement on back of grant)

Delivered to the within named Samuel Hopkins this fourth day of August 1790.

Th: Jefferson

First United States Patent Grant
July 31, 1790
(Reproduced from the original in the collection of the Chicago Historical Society)

Courtesy Chicago Historical Society

Figure 1-1. The first U.S. Patent.

from the adoption of this law, and U.S. Patent 5,000,000 issued on March 19, 1991, to L. O. Ingram et al.

The Patent Act of 1870 was adopted to codify case law that had been created by court decisions up to that time, and in 1887 the United States joined with many other countries in the Paris Convention, which provided for certain limited recognition of the priority rights of inventors from each participating country. The present patent law of the United States is the Patent Act of 1952, the most important new feature being a workable definition of what constitutes invention.

Other Modern Patent Systems

Most countries have patent systems. The major countries have "examination" systems—that is, they undertake to compare applications with previous knowledge (the "prior art") before granting patents. Smaller countries commonly use registration systems, in which patents are automatically granted if certain formalities are met, and enforcement of the patentee's rights is left to the courts. In addition, many smaller countries grant patents of confirmation, based on patents previously issued in one of the major countries.

Patents have force only in the individual countries where they have been granted. Thus, an inventor (or the employer, if that is the case) must file a patent application in every country where patent protection is expected to have value. This multiple filing can be a costly undertaking if it is anticipated that the invention may be widely useful, and the wish is frequently expressed for a single patent that would apply worldwide. That may come in time but surely not very soon.

In fact, cooperative patent movements are in various stages of planning and adoption. A Patent Cooperation Treaty that has already been adopted by the United States and many other countries provides for a single application that will be accepted by all signatory countries, with examination by only one examining group. A single patent application serves for the 17 African countries that were formed from the French colonies, known as the Malagasy Union. The four Scandinavian countries are moving (slowly) toward a common patent. Many European countries accept a single application (the European patent system), although each will issue individual patents according to the requests of the applicant and the patent law of each country. A single Common Market patent applies to all the countries of the European Economic Community (EEC). These developments are discussed in more detail in Chapter 15.

Nature and Purpose of Patent Systems

Regardless of the diversity in detailed practice of the many patent systems of the world, they all share the same fundamental goal. That goal is to provide an incentive to each inventor to disclose his or her findings for the long-term benefit of society rather than to attempt to profit from the invention in secret. The incentive is a short-term right (5 to 20 years, depending on the country) to prevent ("exclude" is the technical term) others from using the invention.

A patent will be granted if the inventor fulfills his or her part of the bargain and if the invention has the three basic elements of patentability, which are

1. It is truly new. "Novelty" is the technical term.

2. It is not obvious in the light of what was known before. "Obvious" and "prior art" are the technical terms.
3. It is useful.

The law of each country has its own way of applying these requirements, that is, through administrative or judicial procedures, but the three basic elements are common to all patent systems.

If these three conditions are met, a patent may be granted. The term *granted* is used because a patent is a grant from a government in return for the disclosure by an inventor of how to practice the invention. A bargain is struck between the parties. An inventor does not have a right to a patent unless he or she has fulfilled his or her part of the bargain, and it is not uncommon for patents to be struck down by the courts when it is found that the inventor failed to provide adequate instructions for practicing the invention. The technical term is an "enabling disclosure".

The word "patent" is commonly used as a noun to refer to the grant held by an inventor, but the legal document is actually known as "Letters Patent", meaning a written instrument (letter) open to the public (patent, adj., from the Latin *patens*, present participle of *patere*, to be open), in contrast to "Letters Close", that is, secret or sealed orders. The official copy of a U.S. Letters Patent is an impressive, colorfully ribboned and sealed legal document.

The most common misconception about patents is that a patent applicant has a right to practice the invention when a patent has been granted. This is often not the case. A patent gives its owner only the right to exclude others from practicing the invention. Often, an invention is an improvement on a broader invention that is the subject of another "dominating" patent held by someone else. The manufacture of a patented product may require the use of process steps patented by others. Patents of others may cover the important uses of a patented product. Improper attempts to exploit patents may run afoul of antitrust laws. All these factors must be considered in the actual use of a patent, and they are discussed further in Chapters 9 and 10.

The preeminent consideration of patent systems is the public interest. Court decisions in challenges to patent validity usually hinge on the questions of whether the patentee really made a new, unobvious, and useful contribution to knowledge and, at least as important, whether a candid and complete disclosure has been made of the new findings. The function of the U.S. Patent and Trademark Office in examining a patent application is twofold:

1. to keep known or obvious technology from being unfairly pre-

empted by an applicant who has not really made an inventive contribution

2.   to encourage disclosure of inventions as a stimulus to the continuing development of technology

The success of our U.S. Patent and Trademark Office in achieving this constructive balance between the competitive thrust and the needs of a cooperative society has assuredly been a potent force in the technological preeminence of this country.

2

How
To Read
a Patent

A patent is a stylized document. Many chemists and chemical engineers approach patents gingerly because they are uncomfortable with the unfamiliar language and structure and perhaps because of a belief, especially common in the academic community, that patents are not an entirely reliable source of technical information. An understanding of how patents are structured and the reasons for the way they are written can make it much easier to use this important source of information effectively.

In contrast to a scientific or technical paper, which presumes background knowledge by the reader, a patent must stand on its own. Prior knowledge of the reader is not assumed other than normal skill in the art to which the patent pertains. Each patent is an individual exposition of the problem addressed, the solution to the problem, and of the many opportunities seen by the inventor for elaboration and practical use of his or her findings. Thus, a patent typically includes both factual and speculative information. The reader, by experience, can learn to sort out and to evaluate these aspects of a patent disclosure, but reading a patent must necessarily be approached in a different way from a scientific paper.

For the chemist, reading a patent can be difficult. Information is often stated, revised, and repeated in language that appears to be unnecessarily formal. However, a patent is a legal document. The text that is unnecessary to a scientist may provide important legally sufficient information at a later stage in the patent process.

The Contents of a Patent

Patent disclosures nearly always include certain common elements, usually in the following order:

1997–4/91/0007/$06.00/1
© 1991 American Chemical Society

1. A statement of the field of technology, that is, the *subject*.
2. A discussion of the prior art, that is, background information, and a statement of the *problem* to be solved.
3. Statements of the "objects" of the invention, that is, the *benefits* provided by the inventor's discovery.
4. A summary or "definition" of the invention, that is, the *solution to the Problem* that the invention provides, stated in technical terms.
5. Detailed elaboration of all aspects of the invention as summarized in the definition.
6. A description of the usefulness ("utility") of the invention.
7. Working examples.
8. Claims, the legal description of what has been granted as an exclusive right to the inventor.

Contemporary U.S. patents are often clearly divided into sections covering each of these elements. Older U.S. and many foreign patents may not be laid out so distinctly, but each of these factors is typically addressed to some degree in the disclosure and claims.

U.S. Patent 3,987,074 is illustrated in Figure 2-1. Some patents are shorter than this one, and some are much longer. The record is held by a U.S. patent in the electronics field that was longer than 900 pages.

United States Patent [19]

Haase et al.

[11] **3,987,074**

[45] **Oct. 19, 1976**

[54] **PROCESS FOR THE MANUFACTURE OF VANADYL ALCOHOLATES**

[75] Inventors: **Ranier Haase**, Bokel(Oldenburg); **Arnold Lenz**, Cologne-Stammheim, both of Germany

[73] Assignee: **Dynamit Nobel Aktiengesellschaft**, Troisdorf, Germany

[22] Filed: **Aug. 21, 1974**

[21] Appl. No.: **499,251**

[30] **Foreign Application Priority Data**
Aug. 25, 1973 Germany.......................... 2343056
Dec. 17, 1973 Germany.......................... 2362704

[52] **U.S. Cl.** .... 260/429 R; 252/431 R
[51] **Int. Cl.²** C07F 9/00
[58] **Field of Search** .... 260/429 R

[56] **References Cited**
UNITED STATES PATENTS
3,432,445   3/1969   Asgan .... 260/429 R X
3,455,974   7/1969   Liong Su........................ 260/429 R
3,652,617   3/1972   Termin et al. .... 260/429 R

3,657,295   4/1972   McCoy.......................... 260/429 R

FOREIGN PATENTS OR APPLICATIONS
1,271,641   1/1962   France
1,816,386   2/1970   Germany

OTHER PUBLICATIONS
Bull. Acad. Sci. USSR pp. 899–900 (1957).

Primary Examiner—Helen M. S. Sneed
Attorney, Agent, or Firm—Burgess, Dinklage & Sprung

[57] **ABSTRACT**

A process for the preparation of a vanadyl alcoholate which comprises contacting vanadium pentoxide with an alcohol in the presence of an orthoester of the formula $R' . C(OR'')_3$ wherein:

R' is hydrogen, a straight-chained alkyl group of 1 to 5 carbon atoms or a branched chain alkyl group of 1 to 5 carbon atoms; and
R'' is a straight-chained alkyl group of 1 to 12 carbon atoms, a branched-chain alkyl group of 1 to 12 carbon atoms or phenyl.

12 Claims, No Drawings

Figure 2-1. U.S. Patent 3,987,074. (Continued through page 13.)

3,987,074

1

PROCESS FOR THE MANUFACTURE OF VANADYL ALCOHOLATES

BACKGROUND OF THE INVENTION

1. Field of the Invention

This invention relates to a process for the preparation of vanadyl alcoholates, particularly vanadyl alcoholates of the formula: $O = V(OR)_3 (Y)_n$ wherein Y is $\equiv$ V=O and n is 0 or 1 and when n is O, R is alkyl, cycloalkyl, alkylaryl, arylalkyl, aryl, alkoxyaryl or hydroxyalkyl and when n is 1, R is an alkoxy radical. This invention is particularly addressed to the problem of removing water formed during the preparation of vanadyl alcoholates by reaction of vanadyl pentoxide with an alcohol. This invention is also directed to the preparation of vanadyl alcoholates of high purity.

2. Description of the Prior Art

It is known that vanadium pentoxide can be reacted with alcohols to form the corresponding vanadyl alcoholates according to the following reaction:

$$V_2O_5 + 6 \text{ ROH} \rightleftharpoons 2 \text{ VO(OR)}_3 + 3H_2O$$

Since this reaction is an equilibrium reaction it is necessary, in order to favor the formation of esters, to continually remove the water that forms. The continual removal of the water, however, involves great difficulty.

It is known to remove the water by distilling over the water as it forms in the reaction, together with excess alcohol, into a second reaction vessel containing a substance which absorbs water, such as quicklime for example. At the boiling temperature of the alcohol, and with intensive stirring, the alcohol is dewatered in the second reaction vessel and then distilled back into the first reaction vessel in which the reaction with vanadium pentoxide is taking place. (Cf. German Offenlegungschrift 1,816,386).

This known process, however, is very expensive. For example, the water-removing substance must be constantly replaced or regenerated. Also, constant vigilance is necessary during the reaction to prevent the reaction from stopping prematurely due to the exhaustion of the water-removing substance. Heat must continually be introduced into the mixture captured in the second reaction vessel in order to keep it boiling. This relatively great energy consumption is particularly appreciable when operating on a commercial scale.

Another disadvantage of this known process consists in the fact that it is necessary to operate in the presence of a strongly acid catalyst. Despite the presence of the catalysts, however, it is necessary to reflux the reaction mixture for more than 8 hours as a rule if it is desired to achieve pure ester yields of about 50%. Yields greater that 50% can be achieved only when operating on a laboratory scale. When the procedure is used on a commercial scale it is found that the yields vary from batch to batch. Consequently, the prior-art process does not achieve reproducible results to a sufficient extent.

Operating in the presence of strongly acid catalysts results in additional disadvantages. The amounts of acid added accelerate the reaction of the V_2O_5 with the alcohol to form the ester, and yet they promote the reduction of the pentavalent vanadium to tetravalent vanadium with the cooperation of the corresponding alcohol. The side reaction, however, is not desired.

2

Since the compounds of the strongly acid reaction are used in practice preferably in amounts of 1 to 10 weight percent with reference to the input vanadium, they are not present in only catalytically effective amounts. Rather, they are present in appreciable percentages which constitutes an impurity in the reaction mixture product. Thus, sulfuric acid is usually found in the form of vanadyl sulfate, in which the vanadium is in the oxidative tetravalent stage. Thus, for example, in a solution of vanadium oxytriisopropylate in isopropanol, with a moisture content of 0.2 weight percent the main part of the tetravalent vanadium is found in the form of solid vanadyl sulfate in the unreacted vanadium pentoxide. In the case of higher moisture contents, the vanadyl sulfate is contained increasingly in the reaction mixture and interferes with the processing.

For example, out of 200 g of V_2O_5, 3000 ml. of isopropanol and 10 ml. of concentrated sulfuric acid, one obtains after 3 hours of reaction solution which still contains 0.39 weight percent of water. When the unreacted V_2O_5 is filtered out of the solution and the isopropanol has been removed by evaporation and the ester has been vacuum distilled, 143 g (= 26% yield) of pure vanadium oxytriisopropylate is obtained. An extraordinarily great quantity of 109 g of distillation residue remains as an unwanted by-product probably due to the high percentage of sulfuric acid and the excessively great moisture content of the reaction solution.

If organic sulfonic acids, such as toluenesulfonic and benzenesulfonic acid, are used instead of sulfuric acid, similar phenomena occur. In the preparation of vanadium oxytriisopropylate, voluminous greenish flakes precipitate from the greenish reaction solution during the progressive concentration and the distillative refinement of the ester that follows; these flakes interfere with the distillation and increase the percentage of distillation residue.

Attempts have already been made to use phenol derivatives or weak acids such as boric acid, for example, as catalysts instead of the strong acids. In the reaction of alcohols having up to 4 carbon atoms, however, these catalysts display no activity (French Pat. 1,271,641).

It is also known in the preparation of vanadyl alcoholates to reflux stoichiometric amounts of vanadium pentoxide and an alcohol having 5 to 5 carbons, respectively, in the presence of benzene (with a ratio of alcohol to benzene of 1 : 1.1 to 1 : 1.35 by volume). The water formed at the boiling temperature is removed from the reaction zone as an azeotropic mixture by the addition of benzene and can then be separated as the heavy phase in a water separator.

At a reaction time of 8 to 12 hours the yields amount to only between 10 and 32% with reference to the vanadium pentoxide input. Especially in the preparation of vanadium oxitri-n-butylate, yields of only 26%, for example, have been achieved (cf. Bull. Acad. Sci. USSR 1957, pages 899–900).

In the preparation of vanadium oxitri-n-butylate it has also been proposed to add toluene instead of benzene to the reaction mixture as an extractant in order to increase the yield. Yields of up to about 65% can be achieved by this process (cf. U.S. Pat. No. 3,657,295), but the time required for the reaction is very long, amounting to as much as 24 hours. In this process a mixture of vanadium pentoxide, toluene and n-butanol is heated to ebullition and the water forming in the reaction is distilled together with the toluene as an

Figure 2-1. U.S. Patent 3,987,074. (Continued through page 13.)

3,987,074

3

azeotropic mixture into a water separator. Here a water-rich phase and a hydrocarbon-rich phase are formed. The latter is continuously recycled to the reactor and the water-rich phase is separated at intervals. Using 1.3 to 1.4 times the stoichiometrically required amount of alcohol, the highest yields are achieved in this case, of 65%. According to this patent, the use of a greater excess of alcohol will not achieve any technical effect as regards increasing the yield, increasing the reaction speed, or the like.

The long reaction time (24 hr.) that is required if it is desired to achieve yields of 65% is particularly disadvantageous. Another disadvantage is that very large amounts of toluene must be used (the ratio of toluene to n-butanol is preferably to be from 1 : 1 to 3 : 1 by volume). In the processing of the reaction mixture that follows, the excess solvent — in this case unreacted alcohol together with toluene — is removed by distillation. It has been found disadvantageous that, before the distillate can be used for the next batch, it must first be readjusted to a specific toluene-butanol content, because otherwise reproducible results cannot be achieved. For this purpose, however, complicated procedures are necessary, which make this process uneconomical and render technical scale operation difficult.

To avoid these difficulties in the prior art methods based on the reaction of V_2O_5 with alcohols it has also been proposed that vanadyl alcoholates be prepared by starting out with $VOCl_3$ and reacting it with alkali metal alcoholates or alcohols; this results in yields of 60% (with reference to $VOCl_3$). These known methods, however, have the disadvantage that the reactant $VOCl_3$, which is very sensitive to hydrolysis, has to be prepared in a separate procedure from V_2O_5. Furthermore, undesired vanadium-containing by-products form in this process, as well as alkali chlorides or HCl as reaction products. The hydrochloric acid that forms has to be neutralized with ammonia in an another separate procedure, so that, by and large, the process is a very complicated one. In addition, these prior-art methods result in products which are not entirely chloride-free. The esters prepared in this manner usually have a reduced shelf life, which is indicated by a dark discoloration. Often they are then no longer suitable for use as a component of a polymerization catalyst.

The preparation of vanadyl alcoholates of C_2 to C_4 alcohols by the transesterification of a lower vanadyl alcoholate with a correspondingly higher boiling alcohol is relatively difficult. These transesterification methods require that the starting product be a vanadyl alcoholate of a low alcohol which in turn is supposed to be more easily accessible than the desired vanadyl alcoholate of a higher alcohol. For the preparation of such "low" alcoholates, however, a process like the one described above (on the basis of $VOCl_3$) has hitherto been recommended. In the transesterification processes, losses of yield have always had to be accepted in order to prepare pure esters, so that the problem of manufacturing pure esters with a high space-time yield, by the use of a very simple method, has basically not yet been solved.

It is, therefore, an object of the present invention to provide a process for the preparation of vanadyl alcoholates starting with vanadyl pentoxide. It is a particular object of the present invention to prepare vanadyl alcoholates in high purity and in good yields within a commercially feasible period of time and especially without the use of separate reaction zones, recycling proce-

4

dures and the like. It is a particular object of the present invention to provide a process by which those vanadyl alcoholates which have heretofore proved difficult to prepare can be synthesized in high purity and within commercially feasible period of time with respect to space-time yield.

SUMMARY OF THE INVENTION

The objects of the present invention are provided by a process for the preparation of a vanadyl alcoholate which process comprises contacting vanadium pentoxide with an alcohol in the presence of an orthoester of the formula $R' . C(OR'')_3$ wherein

R' is hydrogen, a straight-chained alkyl group of 1 to 5 carbon atoms or a branched chain alkyl group of 1 to 5 carbon atoms; and

R'' is a straight-chained alkyl group of 1 to 12 carbon atoms, a branched-chain alkyl group of 1 to 12 carbon atoms or phenyl.

The present invention can be considered to be an improvement over the art known process for the preparation of vanadyl alcoholates wherein vanadium pentoxide is reacted with an alcohol. The improvement comprises including in the reaction mixture an orthoester of the formula $R' . C(OR'')_3$ wherein R' and R'' have the previous assigned significance.

It has been found, in accordance with the present invention, that if an orthoester of such formula is included in a reaction mixture containing vanadium pentoxide and an alcohol that the water formed during the reaction is taken up by the orthoester which, in turn, is hydrolyzed to form an ester an an alcohol.

The invention can be more readily understood when reference is made to the general equation for the reaction of vanadium pentoxide and an alcohol which equation is as follows:

$$V_2O_5 + 6 ROH \rightleftharpoons 2 VO (OR)_3 + 3 H_2O \qquad 1.$$

The alcohol is generally employed during this reaction in at least a stoichiometric amount and preferably it is present in excess. The water produced by the reaction would normally present problems in carrying out the process for the water must be removed if the reaction is to proceed to high yields. By including an orthoester of the type described in the reaction mixture, the water is immediately taken up by the orthoester, which, in turn, is hydrolyzed to formic acid esters or to the carboxylic acid ester and the corresponding alcohol, as the case may be, in accordance with the following equation:

$$R' . C(OR'')_3 + H_2O \rightarrow R'COOR'' + 2 R''OH. \qquad 2$$

DESCRIPTION OF SPECIFIC EMBODIMENTS

In a preferred embodiment of the process of the invention, an orthoester is used which upon hydrolysis releases the same alcohol whose vanadyl alcoholate is to be prepared.

Examples of orthoesters are: orthoformic acid esters such as, for example, trimethyl or triethyl orthoformiate triisoamyl orthoformiate, tripropyl orthoformiate, triphenyl orthoformiate and the like, as well as the corresponding orthoesters of acetic acid, propionic acid, butyric acid and the like. In general, these orthoesters are used which are soluble in the particular reaction mixture that is prepared.

The formic acid ester, or the carboxylic acid ester such as acetic acid methyl ester or propionic acid

Figure 2-1. U.S. Patent 3,987,074. (Continued through page 13.)

3,987,074

5

methyl ester or acetic acid ethyl ester as the case may be, may remain in the reaction mixture until the end of the reaction of the V_2O_5, together with the excess alcohol if any, and the inert organic solvent if any has been added; the low-boiling components, however, are usually distilled away in the course of the reaction, preferably through a column, to prevent the temperature of the reaction mixture from dropping below a certain value which is the optimum for the particular reaction.

In the reaction of vanadium pentoxide with alcohol, it is also possible in accordance with the invention to use an orthoester which upon hydrolysis does not release the same alcohol as the one whose vanadyl alcoholate is to be prepared. If the radicals R in the vanadyl alcoholate are to be identical, it is desirable to remove continually from the reaction mixture the low alcohol that is released. It is advantageous in this case to perform the reaction at the boiling temperature of the reaction mixture in a reaction vessel that is preferably equipped with a column. At the top of the column the low alcohol can be removed, in the form of an azeotrope with the formic acid ester or carboxylic acid formed, if desired, and with the inert solvent if any has been added.

In the process of the invention trimethyl orthoformate or triethyl orthoformate is preferred as the orthoester. These orthoesters are easily accessible on the one hand, and on the other hand they are especially well suited for the purpose of rendering harmless the water formed in the reaction of the vanadium pentoxide with the alcohol. When used in the manner described, they are vitually inert in relation to the vanadyl alcoholates that are formed, as are the corresponding formic acid esters which, as described, can be removed from the reaction mixture during or after the reaction. However, it is to be considered that, parallel with the reaction of the trimethyl orthoformate or triethyl orthoformate with the water, a transesterification reaction with the higher (= higher-boiling) alcohol is constantly taking place, although it has no harmful effect on the vanadyl alcoholate that is to be produced.

The alcohol reacting with the vanadium pentoxide in accordance with reaction equation 1 can be, for example, a C_1 to C_{12} alkanol, which can be branched if desired, such as methanol, ethanol, isopropanol, amyl alcohol and its isomers, etc. Longer-chained alkanols, such as lauryl alcohol for example, may also be used. Also, cyclohexanol as well as bifunctional alcohols such as ethylene glycol, may be used. Generally the alcohol has 1–12 carbon atoms.

The following vanadyl alcoholates, for example, may be prepared directly by the method of the invention: vanadyl trimethylate, vanadyl triethylate, vanadyl triisopropylate, vanadyl tri-n-propylate, vanadyl tri-n-butylate, vanadyl trisecondarybutylate, vanadyl tritertiarybutylate, vanadyl triisoamylate, and the vanadyl trialcoholates of the other isomers of amyl alcohol, vanadyl trilaurate, vanadyl tricyclohexanolate, vanadyl trimethylglycolate, vanadyl triphenolate, vanadyl glycolate $(O = V(OCH_2-CH_2O)_3-V=O)$ etc.

The term "vanadyl alcoholate" in the meaning of the invention is to include the corresponding phenolates, and the term "alcohol" is to include the corresponding phenols.

The method of the invention is especially suitable for the manufacture of vanadyl alcoholates on the basis of methanol and ethanol, which have hitherto been able

6

to be prepared directly and in high purity only with difficulty.

In accordance with Equation 1, the alcohol may be used in stoichiometric amounts with respect to the input V_2O_5. Preferably, however, it is added in excess, especially when an orthoester on the basis of a low-boiling alcohol is used in the reaction of V_2O_5 with a higher boiling alcohol. It is desirable to use the water-free reaction components. However, small percentages of water are generally not objectionable since they are intercepted by the orthoester.

In the process of the invention, the orthoester can be added progressively, i.e., accordingly as the reaction progresses, but usually the full amount is added at the beginning of the reaction, and preferably in an excess, but at least in a molar ratio of $1 : 2$ to $1 : 0.5$ with respect to the water that is released in theory in the reaction of vanadium pentoxide with the alcohol that is used.

The vanadium pentoxide used in the method of the invention is preferably in fine powder form. The reaction is performed at elevated temperature, and if desired inert organic compounds are added, such as benzene, toluene or carbon tetrachloride, which serve on the one hand as solvents and on the other hand they serve the purpose of adjusting the boiling point of the reaction mixture to a certain temperature level. Generally speaking, the process is carried out at a temperature C. from the boiling point of the reaction mixture.

The reaction can be carried out at subatmospheric pressure. It can also be carried out at super atmospheric pressures. However, it is preferably carried out at normal atmospheric pressure.

When low-boiling alcohols are used, such as for example the aliphatic C_1 to C_6 alcohols, the reaction is preferably performed at the boiling temperature of the reaction mixture, while the reaction vessel is best equipped with a separating column of, for example, 10 practical trays. The low-boiling mixtures can be taken off at the superimposed column head consisting, among other things, of a reflux condenser and a temperature measuring device.

The reactants are reacted preferably with intensive stirring for several hours, e.g., 8 to 16 hours. When the reaction has ended, the reaction mixture, still hot if desired, can be separated from the unreacted vanadium pentoxide by decanting or filtering.

If the vanadyl alcoholate that has formed is insoluble or poorly soluble in the reaction mixture, a solvent such as carbon tetrachloride, or a hydrocarbon such as benzene, toluene or the like, which will increase the solubility of the alcoholate can be added to the reaction mixture before the reaction begins in some cases or after it has ended in others. After the unreacted vanadium pentoxide has been filtered out, the solution is concentrated, at reduced pressure if desired. In many cases, in which the vanadyl alcoholate is a solid substance, e.g., vanadyl trimethylate, it is concentrated by evaporation to the dry state and the vanadyl alcoholate can be isolated as the residue. If necessary, the vanadyl alcoholate may be refined by recrystallization or sublimation in vacuo.

In cases in which a vanadyl alcoholate that is fluid is prepared, it is desirable to subject the filtrate to a fractional distillation at reduced pressure, in which case overheating of the wall of the flask is to be avoided and provision must be made for thorough mixing of the contents of the flask, since the vanadyl alcoholates are

Figure 2-1. U.S. Patent 3,987,074. (Continued through page 13.)

3,987,074

7

thermally unstable. For the isolation of the pure vanadyl alcoholates the distillation is usually performed with the use of a small column. The distillation residue in the distillative refinement is in most cases extremely small under these circumstances.

The unreacted vanadium pentoxide remaining in the reaction with low-boiling alcohols such as ethanol contains tetravalent vanadium on only a negligible scale. Since the formation of tetravalent vanadium is largely proportional to the temperature of the reaction mixture, the addition of an inert organic solvent serves to reduce the temperature in the reaction mixture, especially in the preparation of vanadyl alcoholates of higher-boiling alcohols.

The vanadyl alcoholates prepared by the method of the invention are very pure having a purity of at least 95% by weight, especially 99%. They are anion-free and stable in storage.

On account of their high degree of purity they are especially well suited for use as catalysts, e.g., as a component of Ziegler catalysts for the polymerization especially of ethylene with propylene. They may also be used in many other reactions, as for example in condensation polymerizations.

The unreacted vanadium pentoxide is relatively pure in the process of the invention and it is not necessary to regenerate it after each batch. Instead, it may be used repeatedly for additional reactions.

Afterward it is advantageous to regenerate the vanadium pentoxide that remains after several reactions. By thermal treatment at preferably about 300° to 400° C. under the influence of air or oxygen, the unreacted vanadium pentoxide which has meantime become somewhat sluggish can be restored to full reactivity for the reaction of the invention.

In order to more fully illustrate the nature of the invention and the manner of practicing the same, the following Examples are presented:

EXAMPLE 1

In a three-necked flask (capacity 6 liters) provided with a reflux condenser, 200 g (2.75 moles) of V_2O_5, 2500 ml. (62 moles) of methanol and 655 g (6.2 moles) of trimethyl orthoformate were placed. The mixture was heated for 8 hours at ebullition with constant stirring. Upon cooling, yellow crystals of vanadyl trimethylate formed on the walls of the flask. To separate the vanadyl trimethylate from the unreacted vanadium pentoxide, the flask contents were filtered in the hot state in which most of the vanadyl trimethylate is in solution. The filtrate was freed of excess methanol and trimethyl orthoformiate at reduced pressure and elevated temperature. 218 g of vanadyl trimethylate was obtained in the form of yellow crystals.

Yield: 25% with reference to the vanadium pentoxide charged.

EXAMPLE 2

500 g of vanadium pentoxide (2.75 moles), 2100 ml. of ethanol (36 moles) and 734 g of triethyl orthoformiate (5 moles) were placed in a three-necked flask of a capacity of 6 liters. The mixture was heated at ebullition with stirring. The three-necked flask was equipped with a separating column (10 practical trays) and a superimposed column head.

At the column head a mixture boiling at 53° C. was removed, which was an azeotrope consisting of formic acid ethyl ester and ethyl alcohol. Later, as more of the

8

easier boiling component was removed, the temperature at the head of the column rose to 77° C.

After 16 hours of heating the contents of the flask was processed. The unreacted vanadium pentoxide was filtered out and the filtrate was first freed of excess ethanol at reduced pressure and elevated temperature. By means of a separating column (5 practical trays) the separation of the triethyl orthoformiate from the pure vanadyl triethylate (95° C/12 Torr) was performed at 42° C./12 Torr.

The yield was 38% with reference to the vanadium pentoxide charged.

The ratio of pure ester to distillation residue was 43 : 1 (in parts by weight). This calculates to a purity of about 98%.

EXAMPLE 3

The procedure was similar to Example 2, but the reaction was performed with the addition of trimethyl orthoformiate instead of triethyl orthoformiate. Thus 500 g of vanadium pentoxide (2.75 moles) was heated at ebulliation with 3100 ml. of ethanol (53 moles) and 655 g of trimethyl orthoformiate (6.2 moles). The light-boiling substance was again removed at the head of the column starting at about 50° C. and up to about 76° C. After 16 hours of reaction time the mixture was processed as in Example 2.

Yield: 48%

The ratio of pure ester to distillation residue was 53 : 1 (parts by weight). This calculates to a purity of about 98%.

EXAMPLE 4

In an apparatus like the one in Example 2, 500 g. V_2O_5 (2.75 moles) was heated at ebullition with 3100 ml. isopropanol (41 moles) and 873 g trimethyl orthoformiate (8.25 moles). At the top of the column an azeotropic mixture of formic acid isopropyl ester and methanol was removed at 57° C., and later only methanol was taken out. After 16 hours of reaction the processing and distillative refinement were performed as described in the previous examples.

Yield of vanadyl triisopropylate: 48%.

The ratio of pure ester to distillation residue was 100 : 1 (parts by weight). This calculates to a purity of 99%.

EXAMPLE 5

As in Example 2, 500 g of vanadium pentoxide (2.75 moles), 2500 ml. of n-butanol (27 moles) and 916 g of triethylorthoformiate (6.2 moles) were heated at ebullition. An azeotropic mixture of formic acid ethyl ester and ethanol was taken out at the head of the column at 55° C. to 78° C., but later only ethanol was taken out; still later the temperature at the head of the column rose to 107° C. After 8 hours the mixture was processed and pure vanadyl tri-n-butylate was obtained in a yield of 53%.

The ratio of pure ester to distillation residue was 82 : 1 (parts by weight). This calculates to a purity of about 98.8%.

EXAMPLE 6

In a three-necked flask of 6 liters capacity 500 g of vanadium pentoxide (2.75 moles), 3100 ml. of ethanol (53 moles) and 740 g of acetic acid orthomethyl ester (6.8 moles were placed. The mixture was heated at ebullition with stirring. The three-necked flask was

Figure 2-1. U.S. Patent 3,987,074. (Continued through page 13.)

3,987,074

9	10

equipped with a separating column (10 practical trays) and a superimposed column head.

At the beginning, a mixture boiling at 57° C. was taken out, consisting of acetic acid methyl ester and methanol. Later, as removal of the easier boiling components continued, the temperature at the head of the column rose to 74° C. After 16 hours of heating the contents of the flask were processed. The unreacted vanadium pentoxide was filtered out and the filtrate was first freed of the excess ethanol at reduced pressure and elevated temperature. The separation of the excess acetic acid orthoethyl ester formed during the reaction was performed by means of a separating column (5 practical trays) at 42° C/12 Torr from the pure vanadyl triethylate (95° C/12 Torr).

The yield amounted to 40.5% with reference to the vanadium pentoxide.

The ratio of pure ester to distillation residue was 21 : 1.

EXAMPLE 7

In an apparatus like that of Example 1, 500 g of vanadium pentoxide (2.75 moles) was heated at ebullition with 3100 ml. of isopropanol (41 moles) and 740 g of acetic acid orthomethyl ester (6.8 moles). At the beginning an easier boiling fraction was taken out at the head of the column at 57° C; later, as removal continued, the boiling point at the head increased to 80° C. In all, 1.2 liters of distillate were removed. After 16 hours of reaction the processing and refinement by distillation were performed as in the preceding example. Yield: 40.5% of pure vanadyl triisopropylate. The ratio of pure ester to distillation residue was 54 : 1.

EXAMPLE 8

As in Example 1, 500 g of vanadium pentoxide (2.75 moles), 3100 ml. of n-butanol (33.5 moles) and 740 g of acetic acid orthomethyl ester (6.8 moles) were heated at ebullition. At the beginning an easier-boiling fraction was removed at the column head at approximately 57° C. As removal continued the temperature at the head rose over a period of 8 hours to 115° C. Then the mixture was processed as in the above examples. Very pure vanadyl n-butylate was obtained in a yield of 57%. The ratio of pure ester to distillation residue was 27 : 1.

EXAMPLE 9

In an apparatus similar to that of Example 1, 500 g of vanadium pentoxide (2.75 moles) was heated at ebullition with 3100 ml. of ethanol (53 moles) and 1000 g of acetic acid orthoethyl ester (6.7 moles). Soon an easier-boiling fraction could be taken out at the top at 72° C. and consisted substantially of acetic acid ethyl ester. After 16 hours of reaction time, 1 liter of distillate had been taken out and the temperature at the column head had risen to 76° C. Then the mixture was processed as in Example 1. Pure vanadyl ethylate was obtained in a

yield of 33%. The ratio of pure ester to distillation residue was 28 : 1.

What is claimed is:

1. A process for the preparation of a vanadyl alcoholate which comprises contacting vanadium pentoxide with a cyclic or acyclic mono or bifunctional alcohol of 1 to 12 carbon atoms in the presence of an orthoester of the formula

$$R' . C(OR'')_3$$

wherein

R' is hydrogen, a straight-chained alkyl group of 1 to 5 carbon atoms or a branched-chain alkyl group of 1 to 5 carbon atoms; and R'' is a straight-chained alkyl group of 1 to 12 carbon atoms or a branched-chain alkyl group of 1 to 12 carbon atoms or phenyl.

2. A process according to claim 1 wherein the process is carried out in the presence of an inert solvent.

3. A process according to claim 1 wherein the process is carried out at the boiling point of the reaction mixture.

4. A process according to claim 1 wherein said orthoester is trimethyl orthoformiate, triethyl orthoformiate, triphenyl orthoformiate, triamyl orthoformiate, triisoamyl orthoformiate, tri-tert.-amyl orthoformiate, tripropyl orthoformiate, trimethyl orthoacetate, triethyl orthoacetate, triphenyl orthoacetate, triamyl orthoacetate, tri-tert.-amyl orthoacetate, triisoamyl orthoacetate, tripropyl orthoacetate, trimethyl orthopropionate, triethyl orthoproprionate, triphenyl orthopropionate, triisoamyl orthoproprionate, tripropyl orthoproprionate, trimethyl orthobutyrate, triethyl orthobutyrate, triphenyl orthobutyrate or tripropyl orthopropionate.

5. A process according to claim 1 wherein said alcohol is a $C_1 – C_{12}$ alkanol.

6. A process according to claim 1 wherein said alcohol is methanol, ethanol, propanol, isopropanol, n-butanol, sec.-butanol, tert.-butanol, amyl alcohol, an isomer of amyl alcohol, cyclohexanol, lauryl alcohol, phenol or ethylene glycol.

7. A process according to claim 1 wherein the orthoester is introduced progressively to the reaction mixture as water is produced by reaction of said vanadyl pentoxide with said alcohol.

8. A process according to claim 1 wherein the process is carried out for from 8 to 16 hours.

9. A process according to claim 2 wherein said solvent is carbon tetrachloride or a hydrocarbon.

10. A process according to claim 9 wherein said solvent is a hydrocarbon and said hydrocarbon is benzene or toluene.

11. A process according to claim 1 wherein said orthoester is present in a molar amount such that the molar ratio of orthoester to water theoretically produced by reaction of the vanadyl pentoxide with alcohol is from 1 : 2 to 1 : 0,5.

12. A vanadyl alcoholate of the formula O = V(OCH$_2$—CH$_2$O$_3$—V=O.

*   *   *   *   *

Figure 2-1. U.S. Patent 3,987,074. Continued.

Bibliographic Information

The first page of modern U.S. patents carries bibliographic information, facts about the examination process, an abstract, and a drawing showing the invention, if necessary. The patent number and date of the grant are in the upper right corner. The title at the head of the left column is usually informative, but not always. Most patent offices will accept very general titles if they are not grossly misleading. Some patent attorneys prefer to make the title noncommittal so that the subject matter is not revealed in circumstances where patents are listed only by title. The Canadian Patent Office is an exception—informative titles are required. U.S. Patent 3,885,792 illustrates the difference. The title is simply "Game Apparatus". The Canadian equivalent to this patent, No. 1,001,695 is entitled "Game Simulating Dealing in Patented Inventions".

After the title, the inventor or inventors are named. U.S. patents must be granted to individuals, who frequently are assigners to their employer, which is shown next as the assignee. The date on which the application was filed and its serial number are shown, followed by a record of previous applications from which this one stemmed, if any and their status, that is, issued, pending, or abandoned. U.S. Patent 3,987,074 (Figure 2-1) is based on two earlier applications filed in Germany. The U.S. Patent and Trademark Office (PTO) and international classification numbers are shown, as well as the U.S. class or classes in which the examiner searched for prior art.

Next is a list of references that the examiner cited as showing the state of the art up to this invention, including earlier U.S. and foreign patents and other literature. The examiner is named and usually the patent attorney, firm, or agent who represented the applicant is also named. The abstract completes the first page. Unfortunately, the PTO makes no systematic effort to control the quality of abstracts, so they cannot be trusted to be complete or informative. In Figure 2-1, the abstract is useful. The abstract is for information only and is not intended to be part of the disclosure.

The small numbers in brackets before each item are internationally accepted code numbers for identifying each piece of information so it can be recognized regardless of the language. These code numbers are especially useful in locating foreign counterparts of a given patent by reference to the priority date—that is, the first filed application on a given invention. This priority date is shown by all later patents on the same application in every country. The complete code is shown in Figure 2-2.

Drawings

Subsequent pages of a U.S. patent carry the drawings, which are necessary to illustrate mechanical devices, graphical data, flow diagrams, electronic circuitry, and the like. Occasionally photographs may be included, for

ICIREPAT Numbers for Identification of Bibliographic Data on the First Page of Patent and Like Documents

The Paris Union Committee for International Cooperation in Information Retrieval Among Patent Offices (ICIREPAT) has approved revisions in INID Codes (ICIREPAT Numbers for Identification of Data) which became effective for use by the countries which apply such codes to their documents on January 1, 1973. A complete list of the Codes, as revised, appears below.

The purpose of INID Codes is to provide a means whereby the various data appearing on the first page of patent and like documents can be identified without knowledge, of the language used and the laws applied.

[10] *Document identification*

  [11] Number of the document

  [19] ICIREPAT country code, or other identification, of the country publishing the document

[20] *Domestic filing data*

  [21] Number(s) assigned to the application(s), e.g. "Numero d'enregistrement national," "Aktenzeichen"

  [22] Date(s) of filing application(s)

  [23] Other date(s) of filing, including exhibition filing date and date of filing complete specification following provisional specification [1]

[30] *Convention priority data* [2]

  [31] Number(s) assigned to priority application(s) [1]

  [32] Date(s) of filing of priority application(s) [1]

  [33] Country (countries) in which priority application(s) was (were) filed [1]

[40] *Date(s) of making available to the public*

  [41] Date of making available to the public by viewing, or copying on request, an unexamined document, on which no grant has taken place on or before the said date [1]

  [42] Date of making available to the public by viewing, or copying on request, an examined document, on which no grant has taken place on or before the said date [1]

  [43] Date of publication by printing or similar process of an unexamined document, on which no grant has taken place on or before the said date [1]

  [44] Date of publication by printing or similar process of an examined document, on which no grant has taken place on or before the said date [1]

[45] Date of publication by printing or similar process of a document, on which grant has taken place on or before the said date

[46] Date of publication by printing or similar process of the claim(s) only of a document [1]

[47] Date of making available to the public by viewing, or copying on request, a document on which grant has taken place on or before the said date [1]

[50] *Technical information*

  [51] International Patent Classification

  [52] Domestic or national classification

  [53] Universal Decimal Classification [1]

  [54] Title of the invention

  [55] Keywords [1]

  [56] List of prior art documents, if separate from descriptive text

  [57] Abstract or claim

  [58] Field of search

[60] *Reference(s) to other legally related domestic document(s)* [3]

  [61] Related by addition(s) [1]

  [62] Related by division(s)

  [63] Related by continuation(s)

  [64] Related by reissue(s)

[70] *Identification of parties concerned with the document*

  [71] Name(s) of applicant(s) [1]

  [72] Name(s) of inventor(s) if known to be such [1]

  [73] Name(s) of grantee(s)

  [74] Name(s) of attorney(s) or agent(s) [1]

  [75] Name(s) of inventor(s) who is(are) also applicant(s)

  [76] Name(s) of inventor(s) who is(are) also applicant(s) and grantee(s)

Codes [75] and [76] are intended primarily for use by countries in which the national laws require that the inventor and applicant are normally the same. In other cases [71] and [72] or [71], [72] and [73] should generally be used.

Notes concerning the application of INID Codes to U.S. patents:
[1] This item is either not applicable to U.S. patents or, if applicable, is either not coded or not assigned this code.
[2] The respective specific data elements within this category are not individually coded. They are printed in a particular format under the caption "Foreign Application Priority Data" which is identified by the INID Code [30].
[3] The specific data applicable to a particular patent is printed under the caption "Related U.S. Application Data." Where the relationship is due solely to division or to continuation and/or continuation-in-part, the data is identified by the appropriate specific INID Code, i.e., [62] or [63], respectively. Where the relationship is due to any combination of these two specific sub-categories, the data is identified by use of the generic INID Code [60].

Figure 2-2. Complete ICIREPAT code for identifying bibliographic data on all patents and similar documents. ICIREPAT stands for international cooperation in information retrieval among examining patent offices.

instance an X-ray birefringence pattern of a crystalline polymer. The patent in Figure 2-1 does not need that sort of illustration. A detailed description of the drawings included is also given in the patent.

The Specification

The specification, or disclosure, follows the drawings. A specification is the inventor's teaching of the invention and how it may be practiced. U.S. patents have no page numbers but are arranged two columns to the page, numbered serially and with numbered lines. This format makes for easy and concise reference to portions of interest.

In U.S. Patent 3,987,074 (Figure 2-1), *the field* in which the invention lies is described in Col. 1, lines 6–17. The remainder of Col. 1, all of Col. 2, and Col. 3 to line 61 discuss earlier methods for making vanadyl alcoholates and the practical difficulties that they present. In other words, this is the statement of *the problem.* From Col. 3, line 62 to Col. 4, line 6, objects of the invention are discussed—that is, *the benefits* from the improved synthesis of vanadyl alcoholates that is the subject of this invention.

Finally, at Col. 4, lines 10–19 is the *definition* of the invention. It is *the solution to the problem* and is the heart of the disclosure. This is the statement to look for when seeking the purpose of a patent. It can be spotted easily in present U.S. patents that use the subheading *Summary of the Invention,* but many patents are not so ordered. Phrases such as "it is the finding of the present invention", "the invention comprises", "in accordance with the present invention", "my invention is characterized in that", are clues to the definition if it is not immediately apparent.

The remainder of the specification, up to the claims, is a detailed explanation and illustration of the terms and limits summarized in the definition and, in general, tell how to practice the invention. In Figure 2-1, Col. 4, lines 20–53 discuss the mechanism of the reaction used in the new process for making vanadyl alcoholates. Column 4, lines 59–66 lists ortho esters that were tested and found useful as well as related ortho esters that the inventors were confident would also be effective. This section is the big difference between a patent disclosure (a legal document) and a scientific paper. The disclosure is "broadened" to include all materials, conditions, and procedures that the applicant and the attorney believe to be equally likely to be "operative" in order to lay the basis for claims that will protect all practical "embodiments" of the invention. This convention is perfectly legitimate in patent practice, although it introduces the risk of disclosing inoperable ways of practicing the invention.

The conditions under which the reaction may be carried out are discussed in Col. 5, lines 10–24. The alcohols that can be used are defined in Col. 5, lines 43–51, and representative vanadyl alcoholates that can be made by this process are described in Col. 5, lines 52–65. Continuing in Cols. 6 and 7, suitable proportions of the reactants, reaction temperatures, pressures, and times are discussed, all in more detail than would be expected in a research publication. Each parameter is described in the broadest operable terms, in the preferred range, and in the "best mode". The best mode must be present, although it need not be specifically identified. All of this detail adds to the tendency of the chemist or engineer to lose patience with the whole document. This detail is there for a reason, however, and once accepted, it is not difficult to locate the technically meaningful information. The requirement that a patented invention be useful is met in this example by the discussion in Col. 7, lines 19–24.

Examples

The remainder of the disclosure is devoted to actual working examples that illustrate as broadly as practical how to practice the invention, how to use the products, and how the invention differs from the prior art. These items can generally be trusted to be accurate and reproducible, although an occasional patent will violate the trust of the patent system by including incomplete or unreliable data. Such patents often cannot be defended and are invalidated by the courts. The great majority of patents contain reliable information because responsible users of the patent system recognize that it is in the interest of all to provide complete and accurate data.

Generic Formulas

Another characteristic of patent disclosures and claims is the frequent use of generic formulas to describe all the compounds of a given type that can be used in a new composition or process or that can be products obtainable by a new process. A typical description is the following, from U.S. Patent 3,987,008.

. . . a polyphosphate having the following formula:

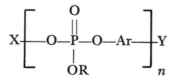

where Ar is a divalent radical having up to about 20 carbon atoms selected from arylene and haloarylene; R is a monovalent radical having up to about 20 carbon atoms selected from alkyl, aryl, haloalkyl, and haloaryl; n is an integer greater than 1; X is a monovalent radical selected from the group consisting of H, alkyl, aryl, haloalkyl, and haloaryl; and Y is a monovalent radical selected from the group consisting of H, OH, and

where X and R are as already given.

This shorthand method is used to provide a complete description of all the compounds that are included in the scope of an invention. In this case, a tremendous number of compounds is included by the generic formula for these polyphosphates, any one of which would be suitable for the purpose of the invention, which was to impart flame-retardant character to polyester compositions.

Legal Terminology

The use of modifiers for physical parameters such as "about" and "at least about" is annoying to literal-minded technical people. Those terms are used very deliberately to avoid sharp cutoffs of protection at the defined limits of ranges of composition, temperature, time, and so forth in patent disclosures and claims. Much trade jargon is used in patent matters, such as the expression "inventive entity" (the inventor or inventors), "conception" (the idea), and "reduction to practice" (carrying out the idea). When these terms have become familiar, they no longer seem strange. This book will try to be helpful in understanding them, and a Glossary is included at the end of the book.

The Claims

The patent concludes with the claims that were allowed by the PTO. These claims are the legal description (the metes and boundaries) of the exclusive right granted to the inventor. The understanding of patent claims is a special problem and is discussed in detail in Chapter 9.

For our purposes here, independent claims are usually the broadest statement of the invention. An independent claim does not have to be broad. Many times a claim to a new compound or a preferred narrowly defined process is written in independent form. Dependent claims refer explicitly to an earlier cited claim (or claims), and incorporate all of the limitations recited in an earlier claim (or claims). Dependent claims are narrower than independent claims and are sometimes referred to as "species" claims, that is, independent claims come first, with the narrower interpretations of the invention in the subsequent dependent claims.

The claims of U.S. Patent 3,987,074 show that the applicants succeeded in convincing the examiner that their improved method for making vanadyl alcoholates as described in the definition at Col. 4 was patentable over the prior art. Claim 12 was also allowed to a particular vanadyl glycolate that had not previously been known. Comparing Claim 12 with the disclosure of this compound at Col. 5, line 61 reveals that a printing **error** was made. Such errors can be corrected by applying to the PTO for a Certificate of Correction, which will be printed and attached to all copies of the patent, as in Figure 2-3.

UNITED STATES PATENT AND TRADEMARK OFFICE
CERTIFICATE OF CORRECTION

PATENT NO. : 3,987,074
DATED : OCTOBER 19, 1976
INVENTOR(S) : RAINER HAASE and ARNOLD LENZ

It is certified that error appears in the above–identified patent and that said Letters Patent are hereby corrected as shown below:

[75] "Ranier" should read -- Rainer --; "Bokel (Oldenburg)" should read -- Bokel/Oldenburg --.

Claim 12, the formula should read -- $O=V(OCH_2-CH_2O)_3-V=O$ --.

Signed and Sealed this

Fifth Day of April 1977

[SEAL]

Attest:

RUTH C. MASON
Attesting Officer

C. MARSHALL DANN
Commissioner of Patents and Trademarks

Figure 2-3. Certificate of Correction for U.S. Patent 3,987,074.

Those who are experienced in reading patents often turn first to the claims to get a sense of the property right represented by the patent. They will then scan the examples to see whether the technical data presented are commensurate with the scope of the claims, and finally they will return to the first sections of the disclosure to round out their understanding of the thinking of the patentee on the breadth of the invention. This procedure is the reverse of the approach used in reading a scientific article, but it is the natural thing to do in view of the great importance of the claims to the business significance of the patent.

Nomenclature

Some chemists may be surprised by the nomenclature used in U.S. Patent 3,987,074 for the vanadium esters made by this new process. The editor of a scientific journal might have insisted that these products be given systematic names such as triethyl orthovanadate and trialkyl orthovanadate rather than informal names such as vanadyl triethylate and vanadyl alcoholate. Patent offices do not edit the terminology used by applicants. A judicially recognized principle in patent practice is that "the inventor is his own lexicographer". Terms used in patent disclosures and claims are taken to have the meaning that the inventor defines for them unless they are misleading or depart outrageously from accepted practice. In Col. 5, lines 63–65 of the patent, it is said that for purposes of this invention alcoholate includes phenolates and alcohol includes phenols. This statement would be most unusual in a scientific paper, but it is acceptable for patent purposes.

Occasionally, inventors and their attorneys coin entirely new words to provide distinct characterizing terms for new arrangements of matter or articles of manufacture, thus adding to the argument for novelty of the invention. Such terms are of course carefully defined in the specification, and the reader should look for discussion of the meaning of the terms used if there is any doubt about their significance. Sometimes such coined words become accepted parts of the language of science and technology. For example, "telomer", "telogen", and "telomerization" were originally devised for patent purposes as terms for defining certain new low-molecular-weight compounds and the process for making them by the addition reaction of polymerizable ethylenic monomers and certain chain-transfer agents. The terms "isotactic", "atactic", and "syndiotactic" were coined to describe the various stereoisomeric forms of polypropylene in patent applications filed by G. Natta and his co-workers.

Foreign Patents

Foreign patents are very much like those of the United States. Some countries publish photographic reproductions of typescripts supplied by the applicant rather than setting the text in type. Some issue patents in the name of the assignee rather than the inventor(s). Some allow or even encourage changes in the disclosure for brevity or to eliminate matter not allowed in claims. (A U.S. specification cannot be changed in any way after filing except to correct errors.) Claiming practice varies substantially with the laws of each country. On the whole, however, the similarities in the patents of the countries of the world are much greater than the differences. The purpose of all patent systems is to encourage the disclosure of new findings and to stimulate others to build on what has been learned.

Problems in Reading a Patent

Problems to the reader of a patent therefore include the presence of more detail than he or she is accustomed to in scientific papers, the broadening of the disclosure beyond actual experimental work, the fact that the best way to practice the invention is not always specifically highlighted (although it must be present in a U.S. patent), and, finally, the legal terminology. A patent is in essence a legal contract between a government as grantor and an inventor as grantee to advance the public welfare, and patent attorneys and agents have found by experience that precision of meaning is best achieved by the use of specialized language that is in some ways quite different from that of everyday usage. For example, the following words have developed special meanings in patent usage:

- "Comprising" has come to be an inclusive term meaning that what is comprised *must* be present, but that other undefined and unlimited components may be present as well.
- "Consisting of", on the other hand, is an exclusive term meaning that *only* the named components may be present.
- "Consisting essentially of" is a frequently used term of intermediate scope usually defined in that the named components must be present, but that additional components (or steps in a process) may be present in minor amounts, provided that they do not substantially alter the properties of the composition or detract from the inventive concept.

Patent disclosures generally avoid any detailed discussion or speculation about reaction mechanisms or the reasons why the invention is effective. This omission is another significant difference between patents and research publications. The reasons for leaving out such discussion are to avoid any limitations on scope of the claims that might be introduced by such an interpretation and also to avoid giving any impression that the invention would have been obvious in the light of theory. A constant problem in the evaluation of the question of obviousness of an invention is that by hindsight an otherwise exemplary discovery may appear to have been an easily accomplished step.

3

Patents
as an
Information Source

Patents are a major lode of technical information that does not get the use it deserves. The bias toward relying almost completely on the journal literature that we learn during our academic years is hard to give up, partly because of the extra effort needed to learn how to read patents effectively, as discussed in Chapter 2, and partly because patents, being individual documents, are not readily available in neatly sorted packages, as are the papers in our many specialized scientific and technical journals.

An important characteristic of patent literature that is not always appreciated by those doing chemical research is that new findings of substantial commercial value are often reported in patents well before they appear in the journal literature. This circumstance is especially true in two of the most important fields of chemistry opened in the past four decades. The very prolific area of isocyanate and polyurethane technology received extensive disclosure in patents during the period from just after World War II until the early 1960s with little or no reporting in the journal literature. Similarly, the discoveries of K. Ziegler, G. Natta, and others in polymerization of olefins with coordination complex catalysts were not discussed in other than the patent literature until about 1960, although the first patent applications were filed in 1953 and were published in some countries shortly thereafter.

Gaining access to the patent literature has two aspects: (1) current awareness, keeping up with what is being published in the field of interest, and (2) retrospective searching, digging out the background information stored in the more than 5 million issued U.S. patents as well as at least as many foreign patents.

Finally, U.S. patents become a reference once issued, as of their filing

1997–4/91/0023/$06.00/1
© 1991 American Chemical Society

date. This fact is extremely important in patent filing strategy and prosecution.

Current Awareness

Current awareness sources include the official bulletins of the national patent offices, the publications of Derwent (London), Chemical Abstracts Service, the abstract publications of certain trade associations, and special patent sections of a number of trade journals. Each has its own usefulness.

The bulletin of the U.S. Patent and Trademark Office is the *Official Gazette*, which is published weekly on Tuesdays, the regular issue date of all U.S. patents, and is available in many public and university libraries. The 1500 to 2000 patents issuing on that day are reported with one or more representative claims in numerical order. About 30% of these are listed under the Chemical patents section, the remainder being in the General and Mechanical and in the Electrical sections. The patents are assigned numbers in such a way that they appear in the order of the U.S. classification system, which makes it possible to look for issuing patents in a given field if one knows the system. This procedure is by no means foolproof, however, and the *Official Gazette* is most useful to professional patent personnel who can afford the time to scan the entire volume.

The other major countries publish similar official bulletins. Some countries make newly issuing patents available only at their patent offices, but this practice is obviously of little help to an individual chemist. Unfortunately, some countries never publish patents they issue or register.

The most useful source of current patent information for those employed by industrial firms that are subscribers is the series of abstract bulletins provided by Derwent Publications, London. Well-written abstracts of all patents from 24 countries are published promptly after issuance, usually within 2 to 3 months, in a variety of arrangements. The chemical field was the original concern of Derwent and still receives the greatest attention. One series publishes all chemical patents in individual "country bulletins" for the United States, Great Britain, France, Belgium, the Netherlands, West Germany, Japan, and the Soviet Union. More useful to the working chemist or engineer are 13 classified "alerting bulletins", which are arranged by subject matter. Regular perusal of the appropriate Derwent bulletins is a painless and certain way to keep abreast of the patent literature in a field of interest.

Derwent also offers a number of other services, such as special formats for particular users, for example, the pharmaceutical industry; abstracts in classified, fileable card form; copies of patents on microfilm; indexes by number and assignee; and concordances of patent equivalents. A computer-searchable version of the Derwent files is available as an on-line service. These services are most commonly used by professional information personnel to provide research scientists, patent liaison people, and patent attorneys with the information they need from the patent literature, as well as alerting them to newly issued foreign patents that may become the subject of opposition procedures, as discussed in Chapter 8.

Chemical Abstracts (CA) is also a useful source of information about chemical patents. *CA* differs from Derwent publications in that its abstracts are not as current—they typically appear about 2 to 3 months after issue. *CA* is more selective than Derwent, publishing abstracts only if they include new chemical information in the judgment of *CA's* editors. *CA* does not repeat abstracts when counterpart (equivalent) patents issue in other countries, as does Derwent, although a concordance identifying equivalent patents is provided in the indexes. *CA* is the source of patent information most accessible to the scientist who does not have available the services of a large organization.

More specialized patent abstract publications are available, for instance those of the American Petroleum Institute (API) and of the British Rubber and Plastics Research Association (RAPRA). Several of the specialized trade journals publish regular feature sections commenting on the significance of newly issuing patents in their particular fields. These journal sections can be valuable sources for those whose interests are concentrated in the commercially active subject areas.

Retrospective Searching

Retrospective searching of the patent literature early in a project becomes important when a researcher undertakes a new project and when consideration is being given to filing a patent application on what appears to be a useful discovery. It is obviously foolish and wasteful to repeat what has already been described by others, and it is not possible to get a valid patent on a finding that is not new or unobvious in the light of prior knowledge. Thus, a thorough understanding of the patent literature in a field of endeavor is fundamental to any fruitful research and development effort.

If the technical person is an employee of one of the large, research-oriented companies, the assistance of a staff of information experts is probably available to aid in planning a search or perhaps to provide professional searches. This relationship is most effective if the research person plays an active role in the study because only he or she can be thoroughly aware of what is needed, and the researcher must incorporate what is learned into the research and development plans. On the other hand, the information professional is best equipped to formulate a searching plan and to deal with the modern computer-based information systems. Cooperation between the research person and the information professional can be extremely productive.

Chemical Abstracts is the most complete and best indexed of the generally available sources of information on chemical patent information in the world. A search of *CA* indexes is the first step that should be undertaken when a new research program is being planned. As the work proceeds, a source such as *CA* or the Derwent publications should be followed regularly to keep aware of newly issuing patents.

When it appears that an invention has been made that should perhaps be the subject of a patent application, it becomes especially important to carry out a thorough search of the patent art. This search is best done in the Search Room of the U.S. Patent and Trademark Office located in Crystal City, Arlington, VA, just outside Washington, DC. The public is welcome to use the search files, which are arranged by subject matter, and an average of 1000 people a day (many of them patent professionals) do so. The patents are classified by product, process, and utility. Often the patents can be searched in a few hours. The patents are arranged in numerical order so that the searcher can see the art emerge through time. Most inventors find it economical to use the services of professional searchers if they do not have an information service available through their own organization. These searchers can be contacted through patent attorneys or agents or hired directly. Searching of the U.S. patent literature may also be performed in a number of libraries, as listed in Table 3-1.

Pergamon ORBIT InfoLine provides a computer search service of U.S. patents based on a combination of IFI files going back to 1950 and on a more sophisticated computer index developed by the Du Pont Company going back to 1964. Many U.S. chemical companies use this service.

Over the past several years, the DIALOG Information Services, Inc. has created an excellent general and specific computer-accessible database as a service to subscribers. Most chemical companies and law firms are subscribers and can perform searches for a nominal charge.

Searches of the foreign patent art are not quite as convenient. The U.S. Patent and Trademark Office has more than 9 million foreign patents. Most of these are arranged numerically, however, and are convenient to use only if the required patent numbers have already been identified by some other

Table 3-1. Libraries That Have Printed Copies of U.S. Patents

State	Name of Library	Telephone
Alabama	Auburn University Libraries	(205) 844-1747
	Birmingham Public Library	(205) 226-3680
Alaska	Anchorage Municipal Libraries	(907) 261-2916
Arizona	Tempe: Noble Library, Arizona State University	(602) 965-7607
Arkansas	Little Rock: Arkansas State Library	(501) 682-2053
California	Irvine: University of California, Irvine Library	(714) 856-7234
	Los Angeles Public Library	(213) 612-3273
	Sacramento: California State Library	(916) 322-4572
	San Diego Public Library	(619) 236-5813
	Sunnyvale Patent Clearinghouse [a]	(408) 730-7290
Colorado	Denver Public Library	(303) 640-8847
Connecticut	New Haven: Science Park Library	(203) 786-5447
Delaware	Newark: University of Delaware Library	(302) 451-2965
Dist. of Columbia	Howard University Libraries	(202) 806-7572
Florida	Fort Lauderdale: Broward County Main Library	(305) 357-7444
	Miami-Dade Public Library	(305) 375-2665
	Orlando: University of Central Florida Libraries	(407) 823-2562
	Tampa: Tampa Campus Library, University of South Florida	(813) 974-2726
Georgia	Atlanta: Price Gilbert Memorial Library, Georgia Institute of Technology	(404) 894-4508
Hawaii	Honolulu: Hawaii State Public Library System	(808) 586-3477
Idaho	Moscow: University of Idaho Library	(208) 885-6235
Illinois	Chicago Public Library	(312) 269-2865
	Springfield: Illinois State Library	(217) 782-5659
Indiana	Indianapolis—Marion County Public Library	(317) 269-1741
Iowa	Des Moines: State Library of Iowa	(515) 281-4118
Kansas	Wichita: Ablah Library, Wichita State University	(316) 689-3155

Continued on next page.

Table 3-1. Libraries That Have Printed Copies of U.S. Patents (Continued)

State	Name of Library	Telephone
Kentucky	Louisville Free Public Library	(502) 561-8617
Louisiana	Baton Rouge: Troy H. Middleton Library, Louisiana State University	(504) 388-2570
Maryland	College Park: Engineering and Physical Sciences Library, University of Maryland	(301) 405-9157
Massachusetts	Amherst: Physical Sciences Library, University of Massachusetts	(413) 545-1370
	Boston Public Library	(617) 536-5400 Ext. 265
Michigan	Ann Arbor: Engineering Transportation Library, University of Michigan	(313) 764-7494
	Detroit Public Library	(313) 833-1450
Minnesota	Minneapolis Public Library & Information Center	(612) 372-6570
Missouri	Kansas City: Linda Hall Library	(816) 363-4600
	St. Louis Public Library	(314) 241-2288 Ext. 390
Montana	Butte: Montana College of Mineral Science and Technology Library	(406) 496-4281
Nebraska	Lincoln: University of Nebraska–Lincoln, Engineering Library	(402) 472-3411
Nevada	Reno: University of Nevada–Reno Library	(702) 784-6579
New Hampshire	Durham: University of New Hampshire Library	(603) 862-1777
New Jersey	Newark Public Library	(201) 733-7782
	Piscataway: Library of Science & Medicine, Rutgers University	(201) 932-2895
New Mexico	Albuquerque: University of New Mexico Library	(505) 277-4412
New York	Albany: New York State Library	(518) 473-4636
	Buffalo and Erie County Public Library	(716) 858-7101
	New York Public Library (The Research Libraries)	(212) 714-8529
North Carolina	Raleigh: D. H. Hill Library, N. C. State University	(919) 737-3280
North Dakota	Grand Forks: Chester Fritz Library, University of North Dakota	(701) 777-4888

State	Library	Phone
Ohio	Cincinnati & Hamilton County Public Library	(513) 369-6936
	Cleveland Public Library	(216) 623-2870
	Columbus: Ohio State University Libraries	(614) 292-6175
	Toledo–Lucas County Public Library	(419) 259-5212
Oklahoma	Stillwater: Oklahoma State University Library	(405) 744-7086
Oregon	Salem: Oregon State Library	(503) 378-4239
Pennsylvania	Philadelphia: Free Library	(215) 686-5331
	Pittsburgh: Carnegie Library of Pittsburgh	(412) 622-3138
	University Park: Pattee Library, Pennsylvania State University	(814) 865-4861
Rhode Island	Providence Public Library	(401) 455-8027
South Carolina	Charleston: Medical University of South Carolina Library	(803) 792-2372
Tennessee	Memphis & Shelby County Public Library and Information Center	(901) 725-8876
	Nashville: Stevenson Science Library, Vanderbilt University	(615) 322-2775
Texas	Austin: McKinney Engineering Library, University of Texas	(512) 471-1610
	College Station: Sterling C. Evans Library, Texas A&M University	(409) 845-2551
	Dallas Public Library	(214) 670-1468
	Houston: The Fondren Library, Rice University	(713) 527-8101 Ext. 2587
Utah	Salt Lake City: Marriott Library, University of Utah	(801) 581-8394
Virginia	Richmond: James Branch Cabell Library, Virginia Commonwealth University	(804) 367-1104
Washington	Seattle: Engineering Library, University of Washington	(206) 543-0740
Wisconsin	Madison: Kurt F. Wendt Library, University of Wisconsin	(608) 262-6845
	Milwaukee Public Library	(414) 278-3247

NOTE: All of these libraries offer CASSIS (Classification and Search Support Information System), which provides direct, on-line access to U.S. Patent and Trademark Office data.
[a]This collection is organized by subject matter.

means. Systematic searches can be arranged through patent agents in each country, who are generally contacted through a U.S. patent attorney. The attorney may also use the Institut International des Brevets (IIB) in the Netherlands, which has a large collection of patents from the major countries and a highly professional staff of search experts.

A new international patent information system, the International Patent Documentation Center (INPADOC), was initiated in Vienna in 1974 by the World Intellectual Property Organization (WIPO). Their service so far is limited to bibliographic information and to identifying equivalent patents ("families") based on the same original application, but they plan to add substantive information services in due course.

Sources of Patent Information

- Chemical Abstracts Service, 2540 Olentangy River Road, P.O. Box 3012, Columbus, OH 43210, (800) 848–6538: Searching Assistance (strategy, etc.), extension 3698; Search Service (fee), extension 3707; Document Retrieval and Delivery (fee), extension 3670.
- Pergamon ORBIT InfoLine, Inc., 8000 Westpark Drive, McLean, VA 22102, (800) 421–7229.
- DIALOG, 3460 Hillview Avenue, Palo Alto, CA 94304; (800) 3–DIALOG; Marketing, extension 1, Customer Service, extension 3.
- INPADOC is represented by Pergamon ORBIT InfoLine, Inc., (800) 421–7229.

Obtaining Copies of Patents

When patents that need further study have been identified through a search of *CA* or any of the other sources mentioned, copies will be needed. If neither the services of a professional information group or an internal patent file is available, several other sources of copies are. The U.S. Patent and Trademark Office supplies copies of U.S. patents for $1.50 each, postage included. Also, collections of U.S. patents are available in a number of libraries, as shown in Table 3-1. Foreign patents can be obtained directly from many of the national patent offices or through professional patent agent firms in the countries of interest. Derwent Publications also has a patent copy service that can meet most needs.

Table 3-2. ICIREPAT Country Identification Code Letters

Code	Country	Code	Country
AR	Argentina	JA	Japan
AU	Australia	LU	Luxembourg
BE	Belgium	MX	Mexico
CA	Canada	NL	The Netherlands
CH	Switzerland	NO	Norway
CS	Czechoslovakia	NZ	New Zealand
DK	Denmark	OE	Austria
DL	Germany (East)	PO	Poland
DT	Germany (West)	PT	Portugal
EI	Ireland	RU	Rumania
ES	Spain	SF	Finland
FR	France	SU	USSR
GB	Great Britain (United Kingdom)	SW	Sweden
HU	Hungary	US	United States
IL	Israel	YU	Yugoslavia
IN	India	ZA	South Africa
IT	Italy		

Citations of patents by number often identify the country by use of the internationally accepted ICIREPAT two-letter country codes shown in Table 3-2. Derwent uses this system with two exceptions. For Japan, Derwent uses only J. For West Germany, they use the standard DT for Offenlegungschriften, or publications of unexamined patent applications, and DS to distinguish Auslegeschriften, examined applications laid open for opposition before final grant. Granted German patents (Patentschriften) would be designated DT although these are not published again on final grant.

Many avenues are open to those who need patent information. It is important that they be understood and used effectively by the chemical practitioner, both to obtain needed background information for research

4

Deciding
Whether To File
a Patent Application

The discovery of something that is new in human experience is one of the greatest satisfactions in the life of a research scientist or engineer. It is sometimes compared to having a child, and includes the normal continued protection of the creation (child) by proper parenting. It is entirely natural that such a person feels a strong proprietary interest in his or her brainchild, and one of the first reactions is "We should patent this!" In many cases that is just what should be done, but a number of factors must be considered before undertaking to file a patent application.

This chapter is written primarily from the point of view of the inventor who is employed by one of the research-oriented companies, institutes, or governmental laboratories. The factors to consider are much the same for the individual working alone or in conjunction with other independent or employed inventors.

Factors To Consider in Deciding To File

The inventor will be expected to answer the following questions when he or she says "We should protect this":

1. What is the advance (the invention, idea, concept, and reduction to practice)?
   a. Should or can the invention be protected as a trade secret and thus not disclosed to the public?
   b. Is the advance proper patentable subject matter under Title 35 of the U.S. Code, Section 101 (35 U.S.C. 101), for example, a new and useful process or composition of matter. (*See* Technical Question 1, "What Is the Invention?")

1997–4/91/0033/$06.00/1
© 1991 American Chemical Society

2. How does it fit into the inventor's or the employer's business interests?

3. What is a reasonable estimate of the economic value of the invention?

4. Are there patents owned by others that might limit the freedom to use the invention?

5. How is the invention superior to alternative products or processes that can serve the same purpose?

6. When was the invention made? Where is the experiment recorded?

7. Have potential U.S. or foreign rights of a patent or a trade secret been damaged, weakened, or destroyed by acts of the inventor(s) or others?

   a. Have verbal or written confidential or nondisclosure agreements been used?

   b. Has the invention been placed on sale?

   c. Has the invention been offered for sale, for example, by means of a verbal or written description or a brochure?

   d. Has an unrestricted sale occurred?

   e. Has there been an enabling oral public presentation or description without any confidential caveat?

   f. Has there been a public presentation of an enabling written or oral scientific paper?

   g. Has the product been tested or used by members of the public, that is, other than the inventors or employees of the organization? Under what written or oral agreements?

   h. Has any experimental use been properly documented?

8. Does the organization have other patents or applications that are related to this invention?

9. What is known about the prior art? Where did the inventor search?

10. How complete are the inventor's data? Will the applicant and organization have to spend more time and money to develop enough information to proceed with a patent application?

These questions are discussed in detail in the next section.

The answers to these questions are all-important to the decision to file, which is essentially a business matter. Not many companies will go to the expense and effort of filing a patent application only to bolster the ego of the inventor. An economic incentive must be present. Supplying the answers to the questions is primarily the responsibility of the inventor, although help can be expected from other groups in the organization in obtaining some of the information needed.

The business decision on whether to file a patent application will depend heavily on the expected uses and value of the patent that may be obtained. If the invention is a potential product or is a process that is likely to be used in actual manufacture, the case for filing an application is readily supported. If the invention is of only speculative value, a more substantial leap of faith is necessary for those who must pay the costs of filing and prosecuting the application. Patent prosecution is the filing, action by the U.S. Patent and Trademark Office (PTO), response, and allowance or rejection steps between the examiner and the patent applicant. More detail is given in Chapter 7.

The prospect of possible income from licensing of the patent is an argument in favor of filing, whereas an invention that would be useful only to others, say a customer, may not be considered a good candidate unless it would help build business. A judgment as to whether it would be possible or practical to enforce the patent may be important. This factor is considered in some detail in Chapters 9 and 10.

Figures 4-1 and 4-2 summarize the creation and perfection of intellectual property rights. Figure 4-1 shows a summary of the pathways by which a U.S. patent process might occur. This process is compared with the time process to protect the intellectual property by trade secret measures.

Invention is usually described as having two parts: concept and reduction to practice. The concept is the new idea that departs from the known art and teaching. Reduction to practice usually includes the actual laboratory or experimental data that shows how the idea differs from the known art. Historically, the U.S. patent system was to benefit and protect the rights of the individual inventor or small business person. Usually these individuals did not have the time or funds to exhaustively perform laboratory experiments to cover each and every aspect of the invention. Accordingly, the inventor was permitted to describe in words experiments (always written in the present tense) that were expected to occur (and did not offend the known art) to better explain and define the scope of the invention. Indeed, it is possible to file a U.S. patent application without performing a single experiment (a paper patent application). "Constructive" reduction to practice occurs at the minute of filing of the patent application with the U.S. Patent and Trademark Office (PTO). In such a constructive application, the inventor is expected to "guess right" in a majority of the written embodiments. Only a few "nonoperative" embodiments are permitted.

Invention (A + B)	U.S. Patent	U.S. Trade Secret
Concept (A)	Idea	Idea
Reduction to practice (B)	Laboratory experimental results If no public disclosure more than 1 year (including offer for sale or sale)	Constructive reduction to practice (paper) occurs on filing of U.S. patent application	Laboratory experimental results
Costs: Actual patent, filing fees, attorney fees	Timely file U.S. (or PCT) patent application Public disclosure permitted		Security: limited-access measures nondisclosure agreements employment agreements fences, guards, locked files coded experiments coded notebooks
Prosecute U.S. patent	Timely prosecutions (10–36 months)		
Issue U.S. patent	Timely U.S. patent issues		(Any disclosure will destroy trade secret claim)
Law applicable	Federal law		State law
Terms of Protection	17 years		Potentially no limit on time Reverse engineering permitted Lawful copying permitted

Figure 4-1. Creation of rights: comparison of aspects of patents or trade secrets.

After the U.S. patent is filed, public disclosure, offer of sale, or sale may occur, and equivalent foreign patent rights must be timely filed for within 1 year under the Paris Convention. Patent prosecution (examination) within the U.S. federal law may take from 10 to 36 months. When the U.S. patent issues, it is enforceable for 17 years.

In a similar manner as shown in Figure 4-1, economically valuable intellectual property may be protected by trade secret methods. The most famous trade secret in U.S. commerce is the formula for Coca-Cola, which has been kept secret for more than 100 years by the use of very elaborate security measures. State law is controlling, and there is potentially no time limit on enforcement. On the other hand, a single unprotected disclosure may destroy the trade secret.

Figure 4-2 describes the acts of the inventor or others that create, perfect, weaken, or destroy U.S. or foreign patent rights. As shown in Figure 4-2, U.S. patent rights are potentially fragile and may be destroyed by unadvised acts by the inventor or others. These acts include, for example, public disclosure, public offer for sale, or public sale more than 1 year before filing the U.S. patent application, withholding pertinent data from the PTO, and other inequitable conduct. A little-known destructive act is the improper filing of a foreign patent application prior to filing a U.S. patent application without first obtaining a U.S. foreign filing license. Immediate petition to the enabling branch of the PTO may result in a valid retroactive license to preserve any U.S. patent rights.

In the foreign patent area, **any** public disclosure or sale prior to filing of a foreign patent application will destroy foreign patent rights unless an earlier U.S. priority patent application has been filed, and the equivalent foreign patent applications are properly and timely filed.

Technical Questions

1. What is the invention?

This question asks for a concise technical statement of what the inventor has found that is "new, unobvious, and useful". If possible, it should be written along the lines of a patent claim, although the legal terms and phraseology are not necessary. It is in essence what will be the definition of the invention as discussed in Chapter 2.

Three classes of subject matter can be protected under the patent law: compositions of matter, processes, and devices. Various terms are applied to these three categories. Compositions include pure compounds as well as mixtures of substances. A process is often called a method, and the term

Invention (A + B)	U.S. Patent Rights	Foreign Patent Rights
Concept (A)	Idea	Idea
	If disclosure—1 year to file U.S. patent	Disclosure—loss of all foreign patent rights
Reduction to practice (B)	Lab experiments	Lab experiments
	Public disclosure Public sale Unrestricted sale	Public disclosure— loss of all foreign patent rights
Procedure: Filing for U.S. patent first	U.S. patent (public disclosure permitted)	U.S. priority 1 year to file for PCT or foreign rights (disclosure permitted)
Filing for foreign patent first	Must obtain U.S. foreign filing license prior to foreign filing	Foreign patent filing or PCT: Chapters I and II
	File U.S. Patent within 1 year	
	Timely prosecution	Foreign patent publication
	Issuance of U.S. patent	Foreign national patent filings

Note: Often foreign patent offices publish the patent application 6 months after they are filed. The PTO keeps the application secret until the patent issues. Thus a trade secret can be maintained until the U.S. patent issues, and then the invention has U.S. patent protection for 17 years.

PCT is Patent Cooperation Treaty

Figure 4-2. Creation and destruction of U.S. or foreign patent rights.

includes a method of use. New uses of known articles or compositions can be patented in this way. Devices include both machines and manufactured articles of any kind. In any of these categories, improvements on existing knowledge can be patentable.

Therefore, the invention must be definable as one of these "statutory" categories of patentable subject matter. Unpatentable subject matter includes intangible ideas ("mental steps"), natural products, and scientific principles. Computer programs cannot be patented per se although they can be an instrumentality of a process (U.S. law is in a state of flux on this subject). The patent examiner will reject out of hand an obviously inoperable claim, such as a perpetual motion machine.

If the invention is a composition, it must be described in terms as precise as possible that show the particular limits of parameters that are critical to operability such as chemical structure, proportions of ingredients, state of purity, molecular weight (of a polymer for instance), melting point, solubility properties, and viscosity. If the invention is a process, all factors essential to its operation must be defined, such as pressure, temperature, residence time, equipment and materials used, and rate and kind of flow Furthermore, the limits beyond which the invention does not work must be made clear to avoid any misconceptions and to avoid claiming an inoperable invention.

Those who make the decision on filing a patent application need a clear statement of the invention so they can visualize what might be claimed. The definition should not be adorned by peripheral discussion, speculation about usefulness, and other language that is unrelated to patentable subject matter. Consideration of value of the invention should be kept separate from the definition itself. Remember, the claims define the property that is the invention. Just as a property deed describes the boundary (metes and bounds) of the land, the claims should state the limits of the invention in clear and precise terms

2. How does the invention fit the business interests?

Company managements are not likely to have much interest in an invention that does not relate to existing business unless it is of such a striking and pioneering character that an opportunity for an entirely new venture is apparent. Some salesmanship may be in order when answering this question.

3. What is the economic value of the invention?

The answer to this question must obviously involve some speculation, but it is usually possible to make a reasonable estimate of how much of a market there might be, what price range can be visualized for a product, how much could be saved by using an improved process, or of what could be gained by licensing the invention. An attempt must be made to deal with this question, no matter how uncertain the figures may be. The benefit of the invention must outweigh the costs of patenting it and of developing it commercially.

4. Are there patents owned by others that might limit the freedom to use the invention?

As discussed in Chapter 1, ownership of a patent confers only the right to exclude others from practicing the invention, not a right to use it. For instance, the invention may be a blend of polymers, one of which is the subject of a patent. It may be an improved process that is still dominated by an earlier patent for applying sprayed coatings. It may be a better control procedure for an injection molding apparatus that is patented. At least a preliminary search of the patent art should be made to uncover any obviously dominating patents. Business management may not be interested in filing an application if there are patents held by others that dominate the new invention.

If the new invention has substantial value even in the face of such a situation, the decision may be to file and to negotiate for a license to the dominating patent. A patent lasts for 17 years, so even if a license under the dominating patent cannot be obtained, the invention can be used when the dominating patent expires. In any case, it is obviously foolish to proceed in ignorance of whether there is freedom to use the invention. Business managers do not like surprises.

5. How is the invention superior to known alternatives? (What is the advantage of the invention?)

A new compound may be an excellent herbicide, but others may be almost as good and also less expensive and perhaps less toxic. A new phosphor may show a pure white emission characteristic, but the trade is satisfied with those already available. A new process may give a better yield but would require a much greater capital investment than the present process. Management will want to be informed of such circumstances.

6. When was the invention made? Where is it recorded?

These facts about the date of conception (when the idea occurred) and the date of reduction to practice (when the idea was actually carried out in a concrete way) need to be known to guide the attorney. They may be vital if the application becomes involved in an interference proceeding, that is, a proceeding instituted within the PTO to determine who is the first inventor(s) when one or more inventors from different organizations apply for a patent on the same invention.

7. Has there been any public disclosure of the invention?

If the answer to this question is yes, the attorney must know the date because such a disclosure may be an absolute bar to obtaining a U.S. patent by filing a patent application more than 1 year later. The attorney will make sure that the application is filed in time to avoid such a bar. The question of what acts amount to public disclosure is highly technical and are best judged by the attorney. In general, commercial use even if not in public view, an offer to sell, unrestricted disclosure even to a limited number of persons, and of course a printed publication amount to public disclosure. An offer for sale or sale of the invention (or if the invention is process, the product of the process) will also raise an absolute time bar to filing a patent application more than 1 year later. If in doubt, the inventor should provide all facts and leave their evaluation to the attorney.

8. Does the organization have any related patents or applications?

The attorney will be able to plan the disclosure and claims to avoid any conflicting relationships if this information is supplied. This information also helps the attorney to position the invention in the technical field.

9. What is the prior art? How thoroughly has it been searched?

This question may be the most important of all. Beyond the fact that it is only good sense to know what has been done previously in one's field of endeavor, a valid patent cannot be obtained in the face of art that discloses the same finding—that is, "anticipates" the invention—or that would make it obvious to those who have ordinary skills and experience in the field. The law presumes knowledge of all that has gone before, and one of the most common grounds for a finding of patent invalidity in litigation is the discovery of pertinent references that the patent examiner did not consider during prosecution of the application. The frustration, embarrassment, and expense of such a situation can be avoided only by thorough familiarity with the prior art.

The person who has the prime responsibility for familiarity with the prior art is obviously the inventor. In a large research organization, the inventor may have the assistance of a professional information group, but the initiative must be the inventor's. There are of course details of evaluation and interpretation of the prior art and its relationship to the scope and timing of the new finding. These points will be evaluated by the attorney; the inventor should be prepared to report the fact situation. The statements on prior art should include a summary of where the search was made—for example, *Chemical Abstracts,* Derwent, internal files, or other search services—covering what period of publication, and so on. Known commercial practice and literature from commercial suppliers should also be included. This information will make it possible for the patent professionals involved to know how thorough the search was and to suggest or arrange further searching if necessary.

It is imperative that the inventor and his or her co-workers in the development of the invention give all of the references and information to the patent attorney for evaluation. The PTO requires inventors and their attorneys to comply with its duty of disclosure requirements (Rule 56). Failure to do so can be used later to invalidate the issued patent.

10. How complete are the data in hand?

Those who must make the decision will need to know whether the proposal to file an application is based on a concept that has not actually been carried out, for instance a computer simulation of a process, a single promising experiment, or a relatively thorough evaluation of the various limits and aspects of the invention. This situation will affect research budgeting and staffing plans and timing of the patent application filing.

Various procedures are used to gather and to present the answers to these basic questions. Many organizations use forms, variously called "invention disclosure", "invention record", "memorandum of invention", or "patent proposal". Some prefer a less structured letter or memorandum that presents the invention and related information in essay style. Whatever the procedure, the questions to be considered are necessarily the same.

Disclosure Document Program of the PTO

After consideration of all these questions, a decision must be made on whether to proceed with filing a patent application. If there are too many negative factors, those who must make this business decision may want to consider alternatives to filing.

The PTO provides a service for inventors: the acceptance and preservation in secret for 2 years of "Disclosure Documents" as evidence of the date(s) of conception of the invention. This program does not diminish the value of the conventional witnessed or notarized records as evidence of conception of an invention. A $6.00 fee (subject to change) is charged. **More important, this program does not provide the discloser with any patent protection.** Additional information is available free from the PTO or at any of the patent libraries listed in Table 3-1.

A Decision Not To File

The decision may be not to file if the invention appears to be of limited commercial value or if the interest is only defensive, that is, to maintain the freedom of the company or its customers to use the invention. Frequently, the company's interest is defensive when the invention relates to a use of a company product that will actually be carried out by customers. The business purpose is to be sure that all customers can use the invention by making it impossible for any one user or a competitor to patent the use and thus preempt that use for him- or herself or customers.

One possible decision, especially with a process improvement, is to take advantage of the invention by operating in secret rather than filing a patent application. A common reason for deciding to operate in secret is the belief that a patent application would necessarily reveal unpatentable information about other aspects of the process in question that is more valuable than a patent to the new improvement. The invention may relate to a process step that, even if patented, might be difficult to enforce or easy for a competitor to avoid by adopting alternative procedures. If the decision is to operate in secret, a hazard is that a competitor may also discover the improvement and obtain a patent that might cause a problem. This risk must be judged on a case-by-case basis. Trade secrets are subject to certain protection under state and federal laws, as discussed further in Chapter 12.

Publication as an Alternative to Patenting

If the decision is not to seek a patent, it may be decided to rely on public disclosure as a defensive measure. Several methods are available for public disclosure. The invention can be published in a scientific journal or trade publication, or a sales bulletin can be distributed to customers interested in the subject matter. Rapid disclosure can be achieved, at modest cost, by publishing a brief description of the invention in *Research Disclosure*, a British journal that is being used increasingly by many companies.

The risk in such disclosures is that in the United States any other person can file a patent application on the same invention within 1 year if it can be shown that he or she indeed made the invention before it was published. An assessment of this risk is part of the decision to disclose rather than to patent a new finding. The Commissioner of Patents and Trademarks can, at the request of a patent applicant, publish a Statutory Invention Registration (commonly referred to as a SIR). The SIR is published without examination, and the applicant waives any right to receive a patent. The SIR is announced in the *Official Gazette* and is printed in the same manner as a U.S. patent. The primary value of a SIR is the establishment of prior art against another application. The advantage of a Statutory Invention Registration over a simple journal article, sales bulletin, or abstract in *Research Disclosure* is that the application can be used as the basis for an interference proceeding if someone else files a similar patent application. In such a case, the defensive applicant can go back to the earliest date of invention that can be shown to prevent a later applicant from obtaining a patent. The waiver of rights cannot be recalled, but the freedom of the public to use the invention can be preserved by this procedure. Interference proceedings are discussed in some detail in Chapter 8.

A Decision To File

When all the questions discussed have been considered, the alternatives to filing have been rejected, and a decision is made to proceed with a patent application, it is time for the inventor to start working with the attorney or agent, possibly with the assistance of a technical liaison person. That process is the subject of Chapters 5 and 6.

5

The Independent Inventor: Obtaining Patent Protection

Those who are employed by large, research-oriented firms or by a federal government department will have in-house legal departments and possibly a technical patent liaison staff who will ensure that the proper steps are taken to obtain patent protection on their inventions. They will have manuals of procedure, forms to fill out, checklists, and reminder systems to make certain that the necessary actions are taken. Those who do not have such a systematic procedure at their disposal often neglect to obtain patents on inventions that could be of substantial value. This chapter is for the independent inventor, the academic person, or the small company. The flow charts in Figures 4-1 and 4-2 in Chapter 4 are also useful.

Filing by an Individual

It is possible and perfectly legal for an individual to file and to prosecute a patent application. Fees amount to about $630, depending on the length of the specification and number of claims. Payment of one-half of these fees ($315) is available for independent inventors, not-for-profit organizations, or for small businesses (those with fewer than 50 employees). These fees are subject to change.

These materials will help the individual applicant:
- Title 37 of the Code of Federal Regulations (CFR) gives the rules and procedures for obtaining patents. It is available from the Superintendent of Documents (U.S. Government Printing Office, Washington, DC) for $7.00.
- A useful book of advice for the independent inventor is *The Inventor's Patent Handbook*, by Stacy Jones (*see* the Bibliography).

1997–4/91/0045/$06.00/1
© 1991 American Chemical Society

- The American Bar Association (Chicago, IL) publishes an informative pamphlet entitled "What Is a Patent?"

The U.S. Patent and Trademark Office (PTO) is sympathetic to individual applications but does not particularly encourage this approach.

Using Professional Assistance

The patent law and the rules are extraordinarily complex. The chance that an amateur will obtain an enforceable patent that properly protects the invention is slim indeed. As with most aspects of our complicated society, the details are best left to a professional. The vast majority of U.S. patent applicants are represented by patent attorneys or by patent agents. Both are registered to represent applicants in practice before the PTO, having passed a special 1-day examination to demonstrate their competence in drafting and prosecuting patent applications and their knowledge of the rules and procedures. Both have a substantial level of technical training and are held to high ethical standards by law and custom. Patent attorneys have law degrees, have passed at least one state (or the Washington, DC) bar examination, and are qualified to represent clients before the courts if patent litigation is necessary.

Patent litigation (usually infringement) is exclusively within the jurisdiction of the U.S. Federal Courts by patent attorneys, but not patent agents. Patent agents handle only the filing and prosecution of patent applications. Because patent procurement is solely federal law, a patent agent is able to move and relocate freely from one state to another and continue to practice as a patent agent without any specific subsequent state examination or licensing procedures.

After deciding to employ such professional assistance, the inventor can choose an attorney or agent from the telephone directory or on the recommendation of others. The PTO provides a list of registered patent attorneys and agents classified by geographical area. In some cities, the local bar association provides a referral service. As in other fields, these professionals tend to specialize in certain technologies, for example, mechanical, chemical, or electrical. An initial telephone call or interview should indicate whether the attorney or agent is prepared to deal with the subject matter at hand.

If an inventor retains an attorney or agent, it is extraordinarily important that the relationship be completely open and candid. The professional can prepare and prosecute an effective patent application only if complete information is provided on the prior art and on all details of experiments that have been carried out. This need for openness is discussed further in Chapter 6. A prior art search will probably be recommended, and this should be done by the inventor or commissioned with a professional searcher as mutually agreed. Communications with a private attorney are **legally privileged** to ensure complete candor in the relationship.

The cost of employing an attorney or agent will of course vary with the complexity of the subject matter and the difficulty of the prosecution. A relatively straightforward case will probably result in legal fees of at least $2500 to $3500 (U.S. filing fee) and is typically higher. Costs will increase substantially if the subject matter is especially complex, if the invention is s small but significant advance in a crowded technical field, or if prosecution of the application turns out to be difficult.

Using Professional Patent Development Firms

Some commercial firms undertake to obtain patents and attempt to license them in return for a percentage of profits that may result. Such firms show all degrees of competence and ethical standards, and the selection of this route to exploiting an invention should be made with great care. Many individuals who think they have great ideas have been fleeced of substantial sums of money by unscrupulous people in this field, with no return and often no patent. The Federal Trade Commission has recently undertaken to control this kind of predatory operation.

Several well-established sources of help are available to help the inventor exploit his or her invention. The Research Corporation of New York City is a nonprofit foundation established by Frederick Gardner Cottrell, inventor of the electrostatic precipitator, as an instrument for turning the rewards of invention to the support of scientific research. This organization works primarily with those in the academic world, undertaking to file and prosecute patent applications on inventions that their evaluation suggests are in the public interest and have a reasonable chance of commercial success. The inventor and the university receive generous proportions of any profits that result from licensing, and the remainder goes into funds used by Research Corporation to support educational programs and a grant program that has provided initial funding for a number of major developments in the academic research community. Similar programs are carried out by the National Research Development Corporation (NRDC) in Great Britain and the Agence Nationale de Valorisation de la Recherche (ANVAR) in France.

In this country, University Patents, Inc., in Stamford, CT, provides a frankly commercial but also reliable patent development service for academic inventors. The University of Oregon manages an Innovation Center funded by the National Science Foundation that provides evaluation, research, development, and technology transfer assistance to independent inventors, and the Universities of Waterloo and Sherebrook in Canada joined in this effort.

Academic Patent Arrangements

Academic scientists and engineers may also have the services of patent development offices in their universities, depending on the particular institution. The oldest is the Wisconsin Alumni Research Foundation (WARF), best known for its successes with the invention of irradiating milk to promote formation of vitamin D and for warfarin rodenticide. The Massachusetts Institute of Technology, Stanford University, Harvard University, the University of California, and other universities have well-organized staffs to handle patent matters. Patent ownership and the financial arrangements vary widely in the academic world, but the trend is to give a significant portion of any earnings from licensed patents to the inventor. Academic researchers sometimes assign rights to their inventions to companies with which they have a consulting relationship. In such cases the company will pay the costs, and its legal staff will handle filing and prosecution of the patent application. Reward to the inventor then is a matter for negotiation between the parties.

Therefore, inventors in the universities and colleges have several effective routes to patent and to exploit their inventions.

Independent Inventors

Individuals and small companies are more likely to have to depend on their own initiative and acumen and on the help of their chosen patent counsel or agents. One route that occasionally succeeds is to offer to sell rights to the invention to one of the larger corporations. However, most corporations will consider unsolicited proposals of this sort only if the person making the offer has already filed a patent application. The first steps toward obtaining patent protection should usually be made before approaching a potential purchaser.

An independent inventor who is confident that he or she has made a discovery of commercial value should make a determined effort to secure patent protection. Some of the best-known inventions of our era came from such individuals. Among the biggest successes have been Chester Carlson's invention of what is now the Xerox process (U.S. Patent 2,297,691) and Edwin Land's inventions of polarizing films (U.S. Patent 2,543,181) and (with his research staff) of the Polaroid Land camera. Integrated circuit and semiconductor technology was significantly advanced by Robert Noyce of Intel Corporation (U.S. Patent 2,981,877). The U.S. genetic engineering and biotechnology industry is based on the pioneering joint work of Stanley Cohen of Stanford University and Herbert Boyer of the University of California at San Francisco (U.S. Patents 4,237,224 and 4,468,464). Charles Townes invented the maser, precursor of the laser, while at Columbia University. Jacob Rabinow has created a host of valuable inventions, including the magnetic clutch, working at the Bureau of Standards and as an independent consultant. Carlson worked with the Battelle Institute in developing his findings, Townes' invention was handled by the Research Corporation, Rabinow worked through his client companies in most cases, and Land formed his own highly successful company. Each found an effective way to exploit his findings. The opportunity for success of independent inventors through the patent system is available to those who have the imagination and drive to make it work.

6

Preparation
of the Patent
Application

In the earlier, simpler days of chemical research, 40 to 50 years ago, it was common for a chemist, either industrial or academic, to write the entire specification for a patent application. After he or she had done this, a patent attorney would make a few changes of a legal nature, write the claims, and file the application in the U.S. Patent and Trademark Office (PTO). It was generally understood that part of the chemist's job was to write the description of his or her findings. As the pace and complexity of research increased, industrial research organizations in many cases found it more efficient to add to their staffs chemists and engineers who served the specialized function of providing liaison between the inventors and the patent attorneys or agents. These liaison persons work with the inventor to assemble all the required information and to get it into the hands of the attorney in appropriate form. In some companies, the inventor is more likely to work directly with company attorneys or retained outside counsel.

Working with the Attorney or Agent

Whatever the organizational arrangement, the objective is to prepare the best possible patent application. The first step after the decision to file is to arrange a conference including the inventor(s), the attorney, and the liaison person, if any, at which the attorney becomes familiar with the invention, the prior art, and the business incentive for filing. Frequently a management representative will also participate. Agreement should be reached at this conference on the scope of claims that are desired to protect the invention and on whether the experimental data in hand are sufficient to

1997–4/91/0051/$06.00/1
© 1991 American Chemical Society

support that scope. It may become apparent that further experiments are necessary to round out the data. There should be agreement on the timing of filing, depending on the sufficiency of the data and on any urgency imposed by the business situation. The filing process must not be unduly delayed, especially in a competitive field. The reasons for not delaying are discussed further in Chapter 8 under the subject of interferences. If the United States goes to a first-to-file system, rapid filing will become imperative.

The attorney will then need the advice of the chemist on the range of compounds, materials, conditions, and so on that should be disclosed to support the broadest reasonable disclosure of how the invention can be practiced. For instance, if a secondary amine is the catalytic agent used in a reaction, the attorney will want a list of representative secondary amines that could be used, even if only two or three have actually been tested. As another example, if a process is carried out at elevated temperature, the attorney will want to know the lowest and highest temperatures that could be used as well as the preferred range of temperatures.

The attorney will ask for working examples. These should be written by the chemist in sufficient detail so that another chemist can reproduce the experiment exactly. The examples must be accurate and complete because a patent claim based on an example that cannot be repeated—that is, is "inoperable"—will not be sustained in a challenge to its validity in court. Valuable patent protection can be lost because of carelessness in transcribing experimental data from the notebook record to patent examples.

An important aspect, particularly of chemical patents, is the use of "paper" examples. In U.S. patent law, a "constructive" reduction to practice occurs the minute a U.S. patent application is filed with the PTO. ("Constructive" in legal terminology usually refers to a legal fiction useful to accomplish a certain result.) Although a researcher would never publish a scientific paper with proposed or expected data, it is common in U.S. patent practice to use paper examples to further explain and define the invention. So long as the paper examples are scientifically reasonable (that is, they do not offend the chemist's chemical sense), the PTO will consider them to have been performed. Often paper examples are written in the present tense, and actual examples are written in the past tense. In a close case, a U.S. patent examiner will request that specific comparative data be produced to distinguish the claimed invention. At this point the inventor will need to conduct the requested comparative experiments within about 3 months, and will need to produce results that confirm his invention in view of the prior art (that is, in the paper examples and in the claims of the invention, the inventor must have guessed right). Most foreign countries require that actual examples be used in describing the parameters of an invention.

The chemical identity of materials used must be known. The PTO will not accept examples in which materials are described by a trademark or proprietary name, because over a period of years the same name may be applied to changing products or the product may become unavailable. The examples should illustrate as completely as possible the range of materials and conditions that can be used. Comparative examples, that is, control experiments, may be needed. A description of test procedures used is often necessary, although reference to standard methods—for example, ASTM tests—may suffice. The attorney may ask for descriptions of further work that illustrates the usefulness of a product.

In the relationship between an inventor and the patent attorney, it is vital that there be complete candor. The duty of candor (under Title 37 of the Code of Federal Regulations, Sections 1.56 and 1.97) has become increasingly important in recent years as evidenced by the larger number of U.S. patents being declared invalid and unenforceable by the U.S. federal courts, because the duty of candor has not been met. The attorney must know, in addition to successful findings, all prior art, commercial practice and testing, and ambiguous or unsuccessful experimental results in order to evaluate the significance of any apparent discrepancies and to determine whether the patent examiner must be made aware of these facts. The withholding of unfavorable findings can result in charges of fraud by the applicant in securing the patent and the invalidation of important patents, even when the withheld data, if it had been submitted, might not have affected the patentability of the claims. The courts treated owners of fraudulently secured patents severely in the 1980s.

Preparing the Patent Application

With the technical information in hand, the attorney, often with the cooperation of a liaison person, will prepare a draft of the patent application and claims. This draft should be reviewed in detail by all concerned (the inventor, co-workers, and managers) to ensure that no factual errors have crept in, that the disclosure is complete, and that the claims cover the invention adequately. After this review, the attorney will prepare the "formal papers", which include the specification and claims and the legal oath or declaration to be signed by the inventor(s) which affirms that he (she, or they) believe(s) that the invention is truly new, that he (she, or they) made the invention, and that it has not been publicly used, placed on sale, or disclosed more than 1 year prior to the date of filing. An inventor's affirmation of these facts by signing the oath or declaration is subject to the perjury laws.

A patent application is the legal document that will be the basis for all that follows in prosecution and issue of the patent. It becomes the heart of the PTO file on the case and will be scrutinized in detail by opposing parties if any interference or patent litigation should take place. Once the application has been filed, the specification cannot be changed except to correct obvious, inadvertent errors. Therefore, it must be as complete and accurate as humanly possible. Any errors, even typographical mistakes, should be changed only at the direction of the attorney. Changes, if necessary, are made by striking out errors with a single line so the original is not obscured, entering the change, with initialing and dating in the margin by the inventor(s) and also by the notary if a sworn oath is used. Signatures should be in permanent ink, preferably blue or black.

Inventorship

When an individual inventor is working with a private attorney or agent in preparing a patent application, there is little question who the inventor is. However, in a larger organization, where cooperative and overlapping projects are being pursued by several chemists and engineers, often with changing assignments as the work proceeds, the determination of the proper inventor or inventors is sometimes a problem. This determination must be based on provable facts rather than unsupported recollections, and egos and organizational relationships should not influence the decision. U.S. patents are granted to individuals, not to organizations, and valid patents cannot be obtained if the inventors are not correctly named.

Determination of inventorship is a legal matter. Only the attorney or patent agent is qualified to make the determination, and the decision should be made on a factual basis. A clear statement from the organization should summarize the complete chronology of the development, including what person or persons had the original idea of the invention ("conception"), what person did the first successful experiment that falls within the scope of the claims ("reduction to practice"), and any other related facts, such as what persons assisted in the experimental work, any significant further improvements in the invention, and so on. A written inventorship summary is often prepared by the liaison person, in consultation with all others involved. The attorney should be supplied with copies of pertinent notebook records, conference minutes, memoranda, reports, etc., that will verify the information provided and that will document the facts if it ever becomes necessary to prove inventorship during litigation. Every person named in the inventorship record should be given a chance to review the reported facts and indicate agreement that they are correct by signing or initialing the document.

The attorney will consider such questions as whether the original suggestion was sufficiently complete to have comprehended all the critical features of the invention as claimed, whether the reduction to practice was carried out independently or as a direct result of the suggestion, and whether the reduction to practice added significantly to the suggestion. He or she will then decide which individuals actually made inventive contributions to the whole and name one or more of them as the inventor or inventors. There is no limit on the number of co-inventors, but if there are more than two or three, it becomes increasingly difficult to assemble a satisfactory inventorship record. The law on joint inventorship is not completely settled, but in general it is expected that each joint inventor shall have made a contribution to at least one claim. If the number of co-inventors is large, this requirement is hard to meet.

Most research chemists and engineers whose careers have been with large organizations have been in the position of being named inventor on developments that were largely completed by others and conversely of having expended great effort on important developments in which others were named as the inventors. If a contributor is determined to be performing routine tests or tasks as specifically directed by another (as a "pair of hands"), usually such a person is not a co-inventor. In other cases a contributor may be acting as a "textbook" and providing information that is already known. Such a person would not be a co-inventor. These factors should not affect the determination of inventorship.

If the written records are incomplete or so carelessly kept that an unequivocal inventorship determination cannot be made, the attorney has no choice but to base the decision on the recollections of those involved. In such a case it may be difficult to sustain the validity of the patent if it should become the subject of a court challenge.

*   *   *

When the properly executed patent application has been filed in the PTO and a filing receipt has been received indicating its safe arrival, several months or longer will elapse before the application reaches its turn with a PTO examiner. The prosecution phase that follows is discussed in the next chapter.

7

Prosecuting
the Patent
Application

This chapter will describe the office procedures and examination process that the patent examiners in the U.S. Patent and Trademark Office (PTO) follow to determine if a patent should be granted. These steps by the patent examiner and timely responses by the applicant are generally called prosecuting the application.

The U.S. Patent and Trademark Office (PTO), Washington, DC 20231, is a large organization actually located in Crystal City, Arlington, VA, which is on the outskirts of Washington, DC. It employs more than 3000 people, half of them examiners and the rest with technical and legal skills. The files include more than 16 million U.S. and foreign patents and more than 5 million other technical publications. The PTO receives more than 100,000 patent applications each year (now more than 45% from foreign applicants), and about 80% of these mature into granted patents. Foreign patent offices of comparable size include those of Japan, the Soviet Union, and the Institut International des Brevets (IIB) in The Hague, which expanded to become the searching office for the European patent and the Common Market patent.

PTO Procedures

Both the filing system and the examination process in the PTO are organized according to a unique classification scheme that has been developed over many years. One page from the *Manual of Classification* is illustrated as Figure 7-1. The manual and its index comprise about 1000 pages, and the classifications are constantly being revised to accommodate new technologies as they develop and patent applications are filed. The Examining

1997–4/91/0057/$06.00/1
© 1991 American Chemical Society

CLASS 260 CHEMISTRY, CARBON COMPOUNDS

MISCELLANEOUS
```
****************************
*CLASS 518 IS AN INTEGRAL....*
*PART OF CLASS 260 AS SHOWN..*
*BY THE POSITION OF THIS.....*
*BOX. THIS CLASS PROVIDES....*
*FOR FISCHER-TROPSCH.........*
*PROCESSES. .................*
****************************
```
```
****************************
*THE 520 SERIES, CLASSES.....*
*520-528, IS AN INTEGRAL.....*
*PART OF CLASS 260 AS SHOWN..*
*BY THE POSITION OF THIS.....*
*BOX. THIS SERIES PROVIDES...*
*FOR SYNTHETIC RESINS. ......*
****************************
```
```
****************************
*CLASS 530 IS AN INTEGRAL....*
*PART OF CLASS 260 AS SHOWN..*
*BY THE POSITION OF THIS.....*
*BOX. THIS CLASS PROVIDES....*
*FOR LIGNINS, PROTEINS.......*
*PEPTIDES, AND NATURAL.......*
*RESINS. ....................*
****************************
```
```
****************************
*CLASS 534 IS AN INTEGRAL....*
*PART OF CLASS 260 AS SHOWN..*
*BY THE POSITION OF THIS.....*
*BOX. THIS CLASS PROVIDES....*
*FOR NOBLE GAS, RADIOACTIVE..*
*METAL, AND RARE EARTH METAL.*
*CONTAINING COMPOUNDS; AND...*
*AZO-TYPE COMPOUNDS. ........*
****************************
```
```
****************************
*CLASS 536 IS AN INTEGRAL....*
*PART OF CLASS 260 AS SHOWN..*
*BY THE POSITION OF THIS.....*
*BOX. THIS CLASS PROVIDES....*
*FOR CARBOHYDRATES OR........*
*DERIVATIVES THEREOF. .......*
****************************
```
```
****************************
*CLASS 540 IS AN INTEGRAL....*
*PART OF CLASS 260 AS SHOWN..*
*BY THE POSITION OF THIS.....*
*BOX. THIS CLASS PROVIDES....*
*FOR HETERO RINGS HAVING.....*
*MORE THAN SIX RING MEMBERS..*
*AND AT LEAST ONE NITROGEN...*
*RING MEMBER, STEROID........*
*COMPOUNDS CONTAINING HETERO.*
*RINGS, AZETIDINONES AND.....*
*FUSED AZETIDINONES (SUCH AS.*
*CEPHALOSPORINS AND..........*
*PENICILLINS). ..............*
****************************
```
```
****************************
*CLASS 544 IS AN INTEGRAL....*
*PART OF CLASS 260 AS SHOWN..*
*BY THE POSITION OF THIS.....*
*BOX. THIS CLASS PROVIDES....*
*FOR SIX-MEMBERED HETERO.....*
*RINGS HAVING TWO OR MORE....*
*HETERO ATOMS OF WHICH AT....*
*LEAST ONE IS NITROGEN. .....*
****************************
```
```
****************************
*CLASS 546 IS AN INTEGRAL....*
*PART OF CLASS 260 AS SHOWN..*
*BY THE POSITION OF THIS.....*
*BOX. THIS CLASS PROVIDES....*
*FOR SIX-MEMBERED HETERO.....*
*RINGS HAVING A SINGLE.......*
*HETERO NITROGEN ATOM. ......*
****************************
```
```
****************************
*CLASS 548 IS AN INTEGRAL....*
*PART OF CLASS 260 AS SHOWN..*
*BY THE POSITION OF THIS.....*
*BOX. THIS CLASS PROVIDES....*
*FOR FIVE-, FOUR- OR THREE-..*
*MEMBERED HETERO RINGS.......*
*HAVING AT LEAST ONE.........*
*NITROGEN ATOM. .............*
****************************
```
```
****************************
*CLASS 549 IS AN INTEGRAL....*
*PART OF CLASS 260 AS SHOWN..*
*BY THE POSITION OF THIS.....*
*BOX. THIS CLASS PROVIDES....*
*FOR HETERO RINGS HAVING.....*
*SULFUR OR OXYGEN AS THEIR...*
*ONLY HETERO ATOMS. .........*
****************************
```
```
****************************
*CLASS 552 IS AN INTEGRAL....*
*PART OF CLASS 260 AS SHOWN..*
*BY THE POSITION OF THIS.....*
*BOX. THIS CLASS PROVIDES....*
*FOR AZIDES,.................*
*TRIPHENYLMETHANES, CERTAIN..*
*NAPHTHACENES, ANTHRONES AND.*
*ANTHRAQUINONES. ............*
****************************
```

350 R	CARBOCYCLIC OR ACYCLIC
350 L	..Leuco dyes
398	.Fats, fatty oils, ester-type waxes or higher fatty acids
398.5	..With preservative
398.6	..Preparing by oxidation of nonaromatic hydrocarbon mixtures
399	..Sulfur containing
400	...Sulfoxy containing
401	....Amino or amido containing
402	....Aryl nucleus containing
402.5	...Nitrogen containing
403	..Phosphorus containing (e.g., lecithin)
404	..Nitrogen containing
404.5	...Poly-nitrogen containing
404.8	..Reaction products with polycarboxylic compounds
405	..Acylated
405.5	..Unsaturation
405.6	..Isomerization
406	..Oxidized
407	..Polymerized
408	..Halogen containing
409	..Hydrogenated
410	..Synthetically produced higher fatty esters
410.5	...With phenols or aromatic alcohols
410.6	...With acyclic polyoxy alcohols
410.7	....Glycerides
410.8	.....With non-higher fatty noncarboxylic acid
410.9 R	...With acyclic monohydric alcohol
410.9 N	.....Unsaturated alcohols
410.9 V	.....Vitamin A acid esters
412	..Recovery or extraction from residues or organic material
412.1	...From fish livers
412.2	...From legumes, nuts or seeds
412.3	....With acidic or basic material
412.4	....Solvent extraction
412.5	...Foots, textile treating liquors, absorbents, sludges or industrial wastes
412.6	...Rendering
412.7	....With acidic or basic material
412.8	...Solvent extraction
413	..Higher fatty acids or their salts
414	...Heavy metal or aluminum containing

Figure 7-1. Sample page from the Manual of Classification.

Corps is organized in three main divisions: chemical, electrical, and mechanical, as shown in Figure 7-2, from the *Official Gazette* for May 7, 1991. The Chemical Division includes four examining groups.

When a patent application is received by the PTO, it is checked to make sure that it meets the "formal" requirements:

1. physical layout as prescribed (margin requirements)
2. naming of the inventor(s)
3. appointment of an attorney or agent
4. signing of the declaration by the inventor(s)
5. quality and physical layout of the drawings
6. payment of the prescribed fees

The date of receipt is made part of the record, and a serial number that is assigned identifies the application during its entire prosecution. Then

PATENT EXAMINING CORPS

JAMES E. DENNY, Acting Assistant Commissioner
STEPHEN G. KUNIN, Acting Deputy Assistant Commissioner

PATENT EXAMINING GROUPS	Phone Number Area Code 703

CHEMICAL EXAMINING GROUPS

GENERAL METALLURGICAL, INORGANIC, PETROLEUM AND ELECTRICAL CHEMISTRY, AND ENGINEERING, GROUP 110—D. E. TALBERT, Director.	308-0661
ORGANIC CHEMISTRY GROUP 120—JOHN F. TERAPANE, JR., Director.	308-1235
SPECIALIZED CHEMICAL INDUSTRIES AND CHEMICAL ENGINEERING, GROUP— 130 BARRY S. RICHMAN, Director	308-0651
HIGH POLYMER CHEMISTRY, PLASTICS, COATING, PHOTOGRAPHY, STOCK MATERIALS AND COMPOSITIONS, GROUP 150—J. O. THOMAS, Director	308-2351
BIOTECHNOLOGY, GROUP 180—EDWARD E. KUBASIEWICZ, Director	308-0196

ELECTRICAL EXAMINING GROUPS

INDUSTRIAL ELECTRONICS, PHYSICS AND RELATED ELEMENTS, GROUP 210—D. G. KELLY, Director	308-1782
SPECIAL LAWS ADMINISTRATION, GROUP 220—ROBERT E. GARRETT, Director.	308-0511
INFORMATION PROCESSING, STORAGE, AND RETRIEVAL, GROUP 230—GERALD GOLDBERG, Director	308-0754
PACKAGES, CLEANING, TEXTILES, AND GEOMETRICAL INSTRUMENTS, GROUP 240—CARLTON CROYLE Director	308-0771
ELECTRONIC AND OPTICAL SYSTEMS AND DEVICES, GROUP 250—JOSEPH J. ROLLA, Director.	308-0956
COMMUNICATIONS, MEASURING, TESTING AND LAMP/DISCHARGE GROUP, GROUP— 260 STEWART LEVY, Acting Director	308-0962
DESIGN, GROUP 290—ROBERT E. GARRETT, Director	308-0511

MECHANICAL EXAMINING GROUPS

HANDLING AND TRANSPORTING MEDIA, GROUP 310—B. R. GRAY, Director	308-1113
MATERIAL SHAPING, ARTICLE MANUFACTURING AND TOOLS, GROUP 320—N. GODICI, Director	308-1148
MECHANICAL TECHNOLOGIES AND HUSBANDRY PERSONAL TREATMENT INFORMATION, GROUP 330—J. J. LOVE, Director	308-0858
SOLAR, HEAT, POWER, AND FLUID ENGINEERING DEVICES, GROUP 340—JOHN KITTLE, Director	308-0861
GENERAL CONSTRUCTIONS, PETROLEUM AND MINING ENGINEERING, GROUP 350— A. L. SMITH, Director	308-0651

Expiration of patents: The patents within the range of numbers indicated below expire during April 1991 except those which may have had their terms curtailed by disclaimer under the provisions of 35 U.S.C. 253. Other patents, issued after the dates of the range of numbers indicated below, may have expired before the full term of 17 years for the same reasons, or have lapsed under the provisions of 35 U.S.C. 151.

Patents	Numbers 3,800,329 to 3,808,603 inclusive
Plant Patents	3,535 to 3,540 inclusive

1126 OG 13

Figure 7-2. Patent Examining Corps, showing the three main divisions: chemical, electrical, and mechanical.

the application is assigned a classification according to the PTO system, and on the basis of this assignment it is forwarded to the examining group responsible for that particular technology. The class assignment is based on the claims rather than on the disclosure, and this practice sometimes produces odd results. For instance, a new polymer that is claimed on the basis of its use for fabricating implantable artificial organs might be assigned to the appropriate group in the mechanical division. An electric arc synthesis of acetylene from hydrocarbons may be assigned to an electrical group. In general, however, examination assignments go to groups that have had long experience with the particular field of the application.

Information Disclosure Statement

If the inventor has not listed all of the prior art or commercial practice that he or she is aware of in the Background section of the application, then the inventor should file an information disclosure statement, which is a listing of the prior art and a short statement of its relevance. Rules 97 and 99 in Title 37 of the Code of Federal Regulations strongly suggest that such a statement be filed with the application or prior to the first action on the merits.

The Examination Process

The examiner who is given responsibility for the application will take it up in its turn. This will in most cases be within a year of filing, and occasionally it will be within 4 or 5 months. The examiner's search is based primarily on patents and literature in his or her own file collections, accumulated over the years rather than the collection in the public search room. The examiner's concern is centered on the claims, as interpreted in the light of the disclosure. The claims are compared with the teachings of the prior art to reach a conclusion as to whether they define an invention that is new and not obvious. Prior patents in the field are the first source studied by the examiner, but the journal and textbook literature is also considered.

The question of whether the invention is new is relatively straightforward. A clear prior disclosure of substantially the same invention is easy to recognize and difficult to argue against. The big problem is the question of obviousness, which involves a considerable degree of judgment. The examiner will frequently combine references to hold that the claimed invention would have been obvious "to one skilled in the art". A typical argument might run as follows. If a compound or composition known from or suggested by one reference is used in a process or article of manufacture known from a second reference, the result would be an obvious, and hence unpatentable, combination.

Other factors considered by the examiner are adequacy and completeness of the disclosure—that is, whether it truly teaches how to practice the invention—and the sufficiency of the claims—that is, whether they include all the limitations inherent in the disclosure and necessary to define how the patent grant is distinct from the prior art.

The Office Action

When the examiner has made a decision, an "office action" or "official action" is issued similar to that illustrated in Figure 7-3. The cover form summarizes the findings, which on the first action are most likely to be a rejection of some or all of the claims or a requirement for election of some of the claims. A requirement for election means that the claims are believed to encompass more than one distinct invention, and the applicant is required to choose one for further prosecution. The claims that are not chosen or elected will be prosecuted later in this application or in a second divisional application.

Figure 7-3b summarizes the reasons for rejection, with an indication of the art on which the decision is based and the relevant portion of the patent law. Three sections of the law will commonly be cited: 102, the requirement for novelty; 103, the nonobviousness requirement; and 112, the requirements for full disclosure and proper claiming. In recent years, the discussion has been given in a running text, not on a specific PTO printed form. Figure 7-3c is a listing of the references cited by the examiner, and there may also be further pages of explanatory discussion by the examiner. Copies of the references are normally supplied by the PTO. The reverse side of the notice of references cited (Figure 7-3d) carries a copy of the pertinent portions of the patent law, Title 35 of the U.S. Code. The inventor may need to study these provisions when questions about the requirements of the law arise.

Occasionally, a patent will be rejected because it is not proper subject matter as defined in Title 35 of the U.S. Code, Section 101 (35 U.S.C. 101). More recently the definition of what is "proper" subject matter has been expanded to include

- a computer program in a chemical process (*Diamond v. Diehr,* Vol. 203 *U.S. Patent Quarterly* page 44 [Supreme Court 1981] or Vol. 450 *U.S. Patent Quarterly* page 175 [1981])

U.S. DEPARTMENT OF COMMERCE
Patent and Trademark Office

Address : COMMISSIONER OF PATENTS AND TRADEMARKS
Washington, D.C. 20231

Paper No _____/_____

⌐John F. Palmer   Art Unit 140⌐

Mailed   April 22, 1976

▼   ▼

03/22/75   100,000   GROUP 140
ROBERT W. JONES ET AL.

JOHN A. HELLER   This is a communication from the Examiner in
1007 MAIN STREET   charge of your application.
PHOENIX, ARIZONA 85203
                                        Commissioner of Patents
                                        and Trademarks

☒ This application has been examined.

☐ Responsive to communication filed_____.   ☐ This action is **made final.**

A SHORTENED STATUTORY PERIOD FOR RESPONSE TO THIS ACTION IS SET TO EXPIRE

_____3_____MONTH(S) _____DAYS FROM THE DATE OF THIS LETTER.

PART I

The following attachments(s) are part of this action:

a. ☒ Notice of References Cited, Form PTO-892.   b. ☐ Notice of Informal Patent Drawing, PTO-948.

c. ☐ Notice of Informal Patent Application,   d. ☐
Form PTO-152.

PART II

Summary of Action

1. ☒ Claims_____1 - 10_____are presented for examination.

2. ☐ Claims_____are allowed.

3. ☐ Claims_____would be allowable if amended as indicated.

4. ☒ Claims____1 - 10_____are rejected.

5. ☐ Claims_____are objected to.

6. ☐ Claims_____are subject to restriction or election requirement.

7. ☐ Claims_____are withdrawn from consideration.

8. ☐ Since this application appears to be in condition for allowance except for formal matters, prosecution as to the
merits is closed in accordance with the practice under Ex parte Quayle, 1935 C.D. 11; 453 OG. 213

9. ☐ Since it appears that a discussion with applicant's representative may result in agreements whereby the application may be placed in condition for allowance, the examiner will telephone the representative within about 2
weeks from the date of this letter.

10. ☐ Receipt is acknowledged of papers under 35 USC 119, which papers have been placed of record in the file.

11. ☐ Applicant's claim for priority based on an application filed in_____on_____
is acknowledged. It is noted, however, that a certified copy as required by 35 USC 119 has not been received.

12. ☐ Other

Figure 7-3a. First page of sample office action, summarizing the findings.

	CLAIMS [1]	REASONS FOR REJECTION [2]	REFERENCES * [3]	INFORMATION IDENTIFICATION AND COMMENTS [4]
PART III		SERIAL NUMBER *100,000*		GROUP ART UNIT *140*
NOTIFICATION OF REJECTION(S) AND/OR OBJECTION(S) (35 USC 132)				

	CLAIMS [1]	REASONS FOR REJECTION [2]	REFERENCES * [3]	INFORMATION IDENTIFICATION AND COMMENTS [4]
1	1 - 3	35 U.S.C. 112		THE LANGUAGE "POLYUNSATURATED ELASTOMER" IS TOO BROAD AND INDEFINITE IN VIEW OF THE DISCLOSURE. SUCH LANGUAGE READS ON POLYESTERS OR POLYURETHANES WHICH HAVE NOT BEEN DISCLOSED.
2	1 - 5	35 U.S.C. 102	A	"A" DISCLOSES REACTING ETHYLENE AND A HIGHER ALPH-OLEFIN IN THE PRESENCE OF COORDINATION CATALYSTS UNDER SAME CONDITIONS RECITED IN CLAIMS.
3	1 - 10	35 U.S.C. 103	A ∨ B + R	OBVIOUS TO EMPLOY THE PARTICULAR NONCONJUGATED DIENES, AS SHOWN IN "B" OR "R", IN THE PROCESS DESCRIBED IN "A".
4				
5				

* Capital letters representing references are identified on accompanying Form PTO-892.
The symbol "v" between letters represents - in view of -.
The symbol "+" or "&" between letters represents - and -.
A slash "/" between letters represents the alternative - or -.

NOTE: Sections 100, 101, 102, 103, and 112 of the Patent Statute (Title 35 of the United States Code) are reproduced on the back of this sheet.

EXAMINER *John F. Palmer* TEL. NO. (703) –557 *2033*

Figure 7-3b. Second page of office action, summarizing reasons for rejection.

FORM PTO—892 (REV. 9-75)	U.S. DEPARTMENT OF COMMERCE PATENT AND TRADEMARK OFFICE		SERIAL NO. 100,000	GROUP ART UNIT 140	ATTACHMENT TO PAPER NUMBER	1
	NOTICE OF REFERENCES CITED		APPLICANT (S)			
			JONES ET AL			

U.S. PATENT DOCUMENTS

*		DOCUMENT NO.	DATE	NAME	CLASS	SUB-CLASS	FILING DATE IF APPROPRIATE
	A	3 0 2 5 4 6 9	6-1972	ARNOLD	260	87	
	B	3 1 8 3 8 3 4	1-1973	LAWSON	245	12	
	C						
	D						
	E						
	F						
	G						
	H						
	I						
	J						
	K						

FOREIGN PATENT DOCUMENTS

*		DOCUMENT NO.	DATE	COUNTRY	NAME	CLASS	SUB-CLASS	PERTINENT SHTS. DWG	PP. SPEC.
	L								
	M								
	N								
	O								
	P								
	Q								

OTHER REFERENCES (Including Author, Title, Date, Pertinent Pages, Etc.)

R	J. REILLY "COORDINATION CATALYSIS" J. AM. CHEM. SOC., 96, 1063-1072 (1970).	
S		
T		
U		

EXAMINER J. F. PALMER	DATE 3-29-76	

* A copy of this reference is not being furnished with this office action.
(See Manual of Patent Examining Procedure, section 707.05 (a).)

Figure 7-3c. Third page of office action, listing references cited by the examiner.

35 U.S.C. 100. Definitions. When used in this title unless the context other—wise indicates —
(a) The term "invention" means invention or discovery.
(b) The term "process" means process, art or method, and includes a new use of a known process, machine, manufacture, composition of matter, or material.
(c) The terms "United States" and "this country" mean the United States of America, its territories and possessions.
(d) The word "patentee" includes not only the patentee to whom the patent was issued but also the successors in title to the patentee.

35 U.S.C. 101. Inventions patentable. Whoever invents or discovers any new and useful process, machine, manufacture, or composition of matter, or any new and useful improvement thereof, may obtain a patent therefor, subject to the conditions and requirements of this title.

35 U.S.C. 102. Conditions for patentability; novelty and loss of right to patent. A person shall be entitled to a patent unless —

(a) the invention was known or used by others in this country, or patented or described in a printed publication is this or a foreign country, before the in—vention thereof by the applicant for patent, or
(b) the invention was patented or described in a printed publication in this or a foreign country or in public use or on sale in this country, more than one year prior to the date of the application for patent in the United States, or
(c) he has abandoned the invention, or
(d) the invention was first patented or cause to be patented by the applicant or his legal representatives or assigns in a foreign country prior to the date of the application for patent in this country on an application filed more than twelve months before the filing of the application in the United States, or
(e) the invention was described in a patent granted on an application for patent by another filed in the United States before the invention thereof by the appli—cant for patent, or
(f) he did not himself invent the subject matter sought to be patented, or
(g) before the applicant's invention thereof the invention was made in this country by another who had not abandoned, suppressed, or concealed it. In determining priority of invention there shall be considered not only the respec—tive dates of conception and reduction to practice of the invention, but also the reasonable diligence of one who was first to conceive and last to reduce to practice, from a time prior to conception by the other.

35 U.S.C. 103. Conditions for patentability; non—obvious subject matter. A patent may not be obtained though the invention is not identically disclosed or described as set forth in section 102 of this title, if the differences between the subject matter sought to be patented and the prior art are such that the subject matter as a whole would have been obvious at the time the invention was made to a person having ordinary skill in the art to which said subject matter pertains. Patentability shall not be negatived by the manner in which the invention was made.

35 U.S.C. 112. Specification. The specification shall contain a written descrip—tion of the invention, and of the manner and process of making and using it, in such full, clear, concise, and exact terms as to enable any person skilled in the art to which it pertains, or with which it is most nearly connected, to make and use the same, and shall set forth the best mode contemplated by the inventor of carrying out his invention.
The specification shall conclude with one or more claims particularly pointing out and distinctly claiming the subject matter which the applicant regards as his invention. A claim may be written in independent or dependent form, and if in de—pendent form, it shall be construed to include all the limitations of the claim incorporated by reference into the dependent claim.
An element in a claim for a combination may be expressed as a means or step for performing a specified function without the recital of structure, material, or acts in support thereof, and such claim shall be construed to cover the corresponding structure, material, or acts described in the specification and equivalents thereof.

Figure 7-3d. Fourth page of office action, citing pertinent portions of the patent law, Title 35 of the U.S. Code.

- the patenting of genetically created new living organisms (*Diamond v. Chakrabarty*, Vol. 206 *U.S. Patent Quarterly* page 193 [Supreme Court 1980] or Vol. 100 page 2204 [Supreme Court 1980])
- the patenting of genetically altered living nonhuman mammals (P. Leder, assigned to Harvard College, U.S. Patent 4,736,8666, issued April 12, 1988)

Figure 7-3b illustrates a rejection of each type. The application seeks claims to a particular process for making copolymers of ethylene, propylene, and a nonconjugated diene. The first rejection—on 35 U.S.C. 112, the disclosure requirement—alleges that the language of the claim is broader than can be supported by the disclosure. The response by the applicant would probably include some modification of the term "polyunsaturated elastomer" to make it clear that certain olefin polymers are made by the claimed process and that polyesters or polyurethanes are not intended to be included in the claim.

The second ground of rejection is on Section 102, the novelty requirement. The disclosure and claims would reveal exactly what is referred to, but it appears that the examiner regards the process of Reference A (Figure 7-3c) as so similar to the invention of this application that the novelty of this new process is open to question. In many cases the response to the office action successfully clarifies the advance made by the inventor by pointing out distinctions that the examiner may not have recognized. Novelty rejections can sometimes be successfully answered by filing an affidavit "swearing back" of the reference, that is affirming that the invention was in fact made before the "effective date" of the reference, if that is the case. This course assumes that the application was filed after the effective date of the cited reference but within 1 year of its publication date. The effective date of a U.S. patent is the filing date of the application, and the date of grant is its publication date. The effective date for other references and non-U.S. patents is the publication date.

The third rejection is based on Section 103, the obviousness section. Apparently the examiner feels that the monomers of the application, which are known from References B and R, could also be used in the process of Reference A even though they are not disclosed in A, and that any good polymer chemist could have surmised the same result. A successful response to this rejection might be an affidavit describing further experiments that show that the nonconjugated dienes of B and R in fact do not give the desired product if used in the process of A. At one time it was felt that a "flash of genius" must be proved to meet the requirements of nonobviousness. Court decisions have tempered that view, with the aid of the Patent Act of 1952, which made the test "obvious at the time the invention was made to a person having ordinary skill in the art to which the invention pertains". The present law thus gives a rational basis for judging nonobviousness of an invention and also discourages application of hindsight in making that judgment by clearly stating that the test should be applied with respect to what was known at the time the invention was made, without taking into account experience in the interim.

The chemical practitioner is warned that legal obviousness under 35 U.S.C. 103 is not straightforward and is a matter to be thoroughly discussed with the patent attorney or agent. The "obvious to try" or "obvious to run" test without guaranteed certainty of the experimental results is not a proper test for "obviousness" under 35 U.S.C. 103. *See* In re Dien, Vol. 152 *U.S. Patent Quarterly* page 550 (Court of Customs and Patent Appeals, 1967).

In compliance with the duty of candor for the inventor(s) and the attorney or agent under Title 37 of the Code of Federal Regulations, Sections 1.56 and 1.97 (37 C.F.R. 1.56, 1.97), it is strongly recommended that statement of art they examined using a PTO form (which is similar to the form of Figure 7-3c) be timely filed (usually within 3 months of the filing date of the application) with the PTO. The examiner initials the form to show that he or she has read and considered the art.

The background information section of the patent application may have mentioned all the pertinent prior art. If not, a good practice during this first phase of the prosecution is for the attorney to inform the PTO of the closest art that was considered during preparation of the application, with a brief discussion of how the invention of the application differs ("can be distinguished") from what is taught in that art. There are several reasons for this practice:

1. It will accelerate the examination process.
2. This demonstration of candor is bound to make a favorable impression and perhaps forestall any unreasonable judgments by the examiner.
3. If this relevant art should be missed by the examiner, it will surely be found by an opponent if the patent is ever litigated.

The courts are especially critical of patents held by those who fail to call known art to the examiner's attention, even if it does not clearly anticipate or make obvious the invention. Therefore, the attorney must be told of all art known to the inventor, so the attorney can decide what must be mentioned. This kind of disclosure is now required by PTO rule and will almost certainly be written into any revision of U.S. patent law that is passed by the Congress. The compliance with this duty of candor cannot be overemphasized.

The office action will be forwarded to the attorney or to the applicant if he or she is acting alone. Three months is normally allowed for a response to the office action, with extensions to 6 months possible. During this time the attorney will want technical advice from the inventor and from others involved on issues such as: (1) possible misconceptions or erroneous conclusions drawn by the examiner and (2) how the claims might be narrowed to meet the examiner's objections without giving up the most valuable aspects of the invention. If the examiner questions the adequacy of the data or the nonobviousness of the invention, the attorney may ask for further experimental evidence to be submitted to the PTO in the form of an affidavit or declaration. This evidence may be from the inventor or from another technical person who has knowledge and experience in the particular subject matter. The attorney will draw up this legal document, and the person who supplied the information will sign it, swearing to its accuracy.

Response to the Office Action

The response, called an "amendment" when changes are made to the application, may include changes in the claims designed to meet the examiner's objections, arguments in reply to the examiner's interpretation of the art cited, perhaps an affidavit as described, and correction of any minor errors in the specification. After the response is filed, the examiner will again consider the application. If it is still held that the claims are unpatentable, a "final" office action is issued. If newly found art that the examiner has not previously considered is cited, the action is usually not final. The only acceptable response to a final action is an amendment or argument that completely meets the examiner's objections. This step is often accomplished by further narrowing of the claims.

Appeals from the Examiner's Decision

If the applicant is convinced that the examiner's objections are unjustified, the case can be submitted to the Board of Patent Appeals, a group of judges who are senior PTO officials. About 10% of all U.S. applications go to appeal. The Board considers arguments of the applicant and the examiner, which must be submitted in a prescribed legal form. It is not unusual for the Board to reverse the examiner's position on the basis of the appeal arguments. Further appeal from an adverse decision by the Board of Patent Appeals can be made to the Court of Appeals for the Federal Circuit (C.A.F.C., a new appellate court established by the Patent Act of 1982) or the District Court for the District of Columbia. Such a step is taken only in cases sufficiently important to justify the legal costs. The Supreme Court on rare occasions has heard a final appeal on a patent application when the Court was convinced that a point of law would be clarified by its decision.

Most patent applications make the grade in the examination process, usually at the examiner level. Meeting with the examiner in an "interview" is often helpful. Patent examiners cannot be experts in the subject matter of every application, and occasionally the examiner will misunderstand the significance of the patent disclosure or of a reference. Such misunderstandings can often be clarified by a face-to-face discussion. The attorney can do this alone if he or she knows all the technical facts, or the inventor or another technical expert may be asked to attend. Such an interview is normally held at the PTO in Arlington, VA, or it may be conducted on the telephone.

Continuation and Divisional Applications

Tactics in prosecuting patent applications sometimes dictate the filing of further applications deriving from the same invention. If the disclosure is substantially unchanged from the original application, a continuation application can be filed; this action provides an additional opportunity for review. If further experimental work, testing, or use has provided new information that can be added to the original disclosure as the basis for additional or broader claims, the second application is called a *continuation-in-part* ("c-i-p"). If the examiner requires restriction of the claims to only one of what are considered to be two or more inventions in the original application, one or more *divisional* applications can be filed, using the same disclosure but with claims to the invention(s) not selected for prosecution in the first application. In each case the later application, which must be filed while the original application is still pending (that is, not issued or abandoned), is given the benefit of the filing date of the original application for priority considerations. The only exception is that claims based on new information ("new matter") in a continuation-in-part application receive only the benefit of the filing date of that c-i-p application.

Sometimes a series of several continuation applications may be filed, especially in a rapidly changing development. Applications filed only to delay issuance of a patent receive a poor reception in the PTO and the courts, and there is little abuse of the continuation process.

Special Aspects of Chemical Patents

Although the patent statutes are the same for all subject matter, court-developed case law has created certain distinct features of U.S. chemical practice. Those in the chemical field should work with attorneys or agents who are specialists in the chemical arts. Certain questions frequently arise about this field, such as the following:

1. Can a new organic compound be patented if closely related compounds—for example, homologs or position isomers—are already known?

    Yes, if it can be shown that the new compound has properties that could not have been predicted from the properties of the known related compounds. This situation is most often seen in the pharmaceutical and agricultural chemical fields because the physiological activity of organic compounds is so sensitive to minor changes in structure.

2. Does the mere mention of a compound in earlier publications preclude a patent claim by a later inventor?

    The answer is a qualified no. At one time it was thought always to be yes, and a common suggestion was that a computer-produced printout of all possible compounds would forever stop the issuance of patents claiming new organic compounds. It has now been established that a disclosure must "put the public in possession of the invention" by actually teaching how to make the compound in order to serve as an anticipating reference. This condition was established by a decision on appeal of an application claiming the compound tetracyanoethylene, which had been mentioned as one of thousands of possible tetrasubstituted ethylenes in a journal article but had not actually been made by the authors of that paper.

3. Would the use of a patented compound as an intermediate that is not isolated in a multistep process infringe a claim to that compound?

    Yes.

4. Can a new use of a known compound be patented?

The answer is a qualified yes. Uses of physiologically active compounds, formulations including known compounds but having an unobvious use, and new physical forms of known compounds such as a crystalline modification or a particle size effective in a new use are examples of patentable inventions using known compounds.

The satisfactory demonstration of utility for new chemical compounds is often a problem in chemical patent applications, especially in the pharmaceutical field when the effectiveness of a new drug in therapy must be shown. Patent attorneys who specialize in chemical practice will be able to advise the inventor on what is needed in such cases.

Other unique features of chemical patents include the frequent use of what is known as Markush language. A Markush structure is a chemical structure in which some atoms and their connections are specified, but others are allowed to vary in some way. A component of a patent claim may be "chosen from the group consisting of A, B, C, or D" where A, B, C, and D are different but each can serve the same function. In other words, A, B, C, or D can each be a compound, an element, a solvent, or a possible substituent or functional group on a chemical structure; with this system, a number of different chemical compounds can be described by one single claim. Markush structures are allowed in patents to protect the invention for sets of related compounds without having to require the inventor to prepare and test each and every possible example. The frequent use of generic formulas to describe a class of related compounds is another special feature of chemical patents that has already been discussed in Chapter 2.

Allowance and Issuance of the Patent

Allowance and Issuance. When the examiner is satisfied that a patentable invention has been properly defined by the claims, nearly always after some amendment and possibly after the filing of continuation applications, a Notice of Allowance is issued. The final fee must be paid within 90 days. The patent is printed and issued within about 3 months after payment of this fee, and on the day of issuance (always a Tuesday) a notice appears in the *Official Gazette*. The official copy of the patent is sent to the owner, who may be the inventor or the assignee. Similarly ribboned and sealed certified copies can be ordered from the PTO for $5.00, and any number of plain copies can be ordered for $1.50 each. (Orders should be sent to the Commissioner of Patents and Trademarks, Box 8, Washington, DC 20231.)

Continuation or Divisional U.S. Patent Application. So long as a U.S. patent is pending, a continuation or a divisional patent application may be timely filed to present additional claims directed to additional aspects of the invention. The original U.S. filing date is preserved so long as all of the requirements of 37 C.F.R. 1.60 or 1.62 are met; for example, at least one inventor must be the same as on the previously filed application and the disclosure must be present in one previous application to support the additional claims.

A divisional U.S. patent application may be timely filed to prosecute those claims earlier determined by the patent examiner to be to a completely separate and distinct invention from the issuing claims in the pending U.S. patent application. The effective term of an issued divisional patent is 17 years.

Reissue Application. Occasionally it is found after a patent has issued that the claims cover more or less than they really should. They may be found to cover too much if additional prior art comes to light, or it may be realized on the basis of reconsideration or further experience that broader claims than those allowed are really justified by the disclosure. It is possible under these circumstances for the patentee or the assignee to obtain reissue of the patent by surrendering the original patent and filing a new application with the same disclosure and new or modified claims. There is further examination, and if no problems are found by the examiner, a reissue patent is granted, and it will expire on the same date that the original patent did. If the claims are narrowed by this process, the reissue can be applied for at any time in the life of the original. If the claims are to be broadened, the application must be filed within two years of the date of grant of the original patent. The reissue will have a different number, in a different series from the original patent (RE series), and the claims are printed in such a way as to make clear the changes from those of the original patent.

Reexamination. The Patent Act of 1984 provided an additional procedure to reassess the patentability of an invention be either the inventor, assignee, or any interested third party. The governmental fees (exclusive of legal fee) to file a reexamination proceeding is presently more than $2000. See 37 C.F.R. 1.501 to 1.570. Often a reexamination is filed by an entity accused of infringing the patent to be reexamined as a defensive measure. Reexamination must be based upon a printed publication. The PTO will make a decision on the basis of the publications and the filing papers whether or not to reexamine the patent. When reexamination or reissue occurs, patent prosecution begins again.

Foreign Patent Applications

Business considerations often dictate the filing of foreign patent applications on inventions made in this country. The laws of most countries make it necessary to work through a local patent agent, and nearly all U.S. patent attorneys maintain relationships with their counterparts in other countries for this purpose. To obtain the protection of the priority date of invention under the international convention, foreign applications must be filed in the Convention countries within 1 year of the first application filing date. This requires a decision on foreign filing in sufficient time to allow applications to be prepared and translations to be made if necessary. The question of foreign filing therefore should be taken up about 6 months after the U.S. application is filed.

U.S. Foreign Filing License. U.S. patent law does not allow a foreign equivalent patent application to be filed earlier than 6 months after the U.S. filing, primarily for U.S. defense and technology transfer considerations, unless a special license is obtained from the PTO. The license is obtained routinely and provided with the Official Filing Receipt (usually within 4 to 6 weeks after U.S. filing). The filing of a U.S. patent application improperly in any foreign country usually results in the voiding of the U.S. patent application or patent. In some instances, a petition for a retroactive foreign filing license can be made and granted by the PTO. An expedited license of a specific scope can usually be obtained within 3–5 days from the PTO by filing of a proper petition, a copy of the U.S. patent application, and the prescribed fee. This procedure allows the government agencies charged with national security time to consider whether the invention is sufficiently vital to national interests that it should be held secret.

Absolute Novelty. A number of foreign countries have laws that do not allow patents to be granted on an invention that has been publicly disclosed or used in any way before the first application is filed. These are called "absolute novelty" countries. Therefore, while only present U.S. and Canadian law allow 1 year to file after disclosure (and Canada will soon become an absolute novelty country), if there is any possibility that patents in other countries will be desired, it is important that the first U.S. application be filed before any journal articles, public announcements, offers to sell, sales, or any commercial use of the invention. The patent attorney should always be consulted on such questions to avoid actions that could jeopardize the right to obtain valid foreign patents on the invention.

Prosecution of patent applications in the major foreign countries is generally similar to that in the United States but with the added complication that all communications need to go through both the U.S. attorney and a foreign counterpart. The laws on patentable subject matter vary. Pharmaceutical products cannot be patented in certain countries, for example, Mexico. Certain countries do not allow patents on foods or beverages, for example, Brazil. Germany requires a convincing showing of "technical advance", meaning that the invention must be a demonstrable improvement over existing practice or knowledge. Some countries do not allow claims to chemical compositions but only to processes for their preparation. British examiners consider only the novelty of the invention with respect to earlier British patents, leaving the question of obviousness to later consideration by the courts if the patent is ever challenged. The most rigorous examination is carried out by the Netherlands, Germany, Japan, and Sweden.

Patent Application Publication. The United States and Canada do not publish any pending patent applications. Most other foreign countries will publish a patent application about 18 months after filing.

Many foreign countries, including Great Britain, Germany, the Netherlands, Sweden, Japan, and Brazil, allow oppositions to the grant of patent claims after examination and publication. Third parties who know of art or prior uses that were not considered by the patent offices of those countries can submit their objections. Further examination and legal arguments ensue, and often previously allowed claims are narrowed or dropped as the result of such an opposition. About one-third of all examined applications that are published in Germany are opposed in this way. The disadvantages of this procedure are the expense and delay involved, but the advantage is that patents finally granted after such a challenge are for practical purposes unassailable for the remainder of their lives.

Many countries have what is known as a registration system. Patents are automatically issued on all applications that meet elementary requirements such as form, appropriate subject matter, and unity of invention, and it is then up to the patent owner to defend these patent rights in the courts of that country if necessary. This system is used in Belgium and Italy, for instance. Some countries also grant patents of confirmation based on the prior issuance of a patent by another country. Many of the countries that grew out of the British Empire have procedures that parallel those of Great Britain, and the African countries that were once French colonies follow the lead of French patent practice. Most likely, such countries will become more autonomous in their patent procedures in time.

Patent Terms

U.S. and Canadian patents are in force for 17 years from the date of grant except for rare special circumstances. The patents of most countries of the world run from the date of filing of the application. Table 7-1 shows the terms of patents in a number of countries. The decision on whether to file foreign applications needs to take into account the shorter patent terms in some countries, the fact that the patents may run from the filing date, the question of enforceability of the patent, and the possibility that final grant may be delayed by an opposition. The average U.S. patent issues 24 months after filing (as of 1990), but the period of pendency varies widely in other countries. It can range from a month or two in the registration countries to several years even for an unopposed application in Germany and the Netherlands. In Japan, unless the applicant specifically requests it, examination will not begin for 5 years. Occasionally an applicant will have the experience of seeing the German patent granted close to or even after the date that it expires as the result of a prolonged opposition.

*   *   *

U.S. patent prosecution, as of June 1991, has no provision comparable with the opposition procedures of foreign countries. The interference provision of U.S. law is of comparable complexity, however, and that is the subject of Chapter 8.

Table 7-1. Patent Terms

Country	Years	From date of
Argentina	5, 10, or 15	grant
Australia	16	filing
Austria	18	publication
Belgium	20	filing
Brazil	15	filing
Canada	20	filing
China (Peoples)	15	filing
China (Republic)	15	filing
Cuba	17	grant
Czechoslovakia	15	filing

Continued on next page.

Table 7-1. Patent Terms (Continued)

Country	Years	From date of
Denmark	17	filing
Finland	17	filing
France	20	filing
Germany (East and West)	18	filing[a]
Great Britain (United Kingdom)	16	filing[b]
Greece	15	filing[a]
Hungary	20	filing
India	14	filing[b]
Iran	5, 10, 15, or 20	filing
Ireland	16	filing[b]
Israel	20	filing
Italy	15	filing
Japan	15	publication[c]
Korea (South)	12	publication[d]
Mexico	10	filing[f]
The Netherlands	20	filing[e]
Norway	17	filing
Poland	15	filing
Romania	15	filing
South Africa	16	filing[b]
Spain	20	grant
Sweden	17	filing
Switzerland	18	filing
Taiwan	15	filing
Turkey	5, 10, or 15	filing
USSR	15	filing
United States	17	grant

NOTE: The values shown presume the timely payment of foreign annuities, taxes, or maintenance fees.

[a] Plus 1 day.

[b] Of complete specification, which must be filed within 1 year of an initial "provisional" application.

[c] From publication for opposition, but not more than 20 years from filing.

[d] But no more than 15 years from filing.

[e] From end of month of filing, or 10 years from end of month of grant, if longer.

[f] 15 years from grant if prior to 1976.

8

Interferences and the Importance of Records

Most of the countries of the world have patent systems based on the principle of "first to file". This means that a patent is granted to the party who first submits an application to the country's patent office, regardless of who actually first made the inventive discovery. Later applicants have no recourse other than to oppose issuance of the patent to the first applicant or to use their prior knowledge as a defense against any charge of infringement if their evidence can be used under the laws of the particular country. The Commission on the Patent System appointed by President Johnson recommended in its 1966 final report on proposed changes in our patent law that the United States should also adopt a first-to-file provision. The American tradition that the patent should go to the person who is the true first inventor is so strongly entrenched, however, that adoption of a first-to-file system has essentially been dropped from consideration by the Congress. Both the United States and Canada have procedures for determining the first inventor, called interference and conflict proceedings, respectively.

Establishment of an Interference

An interference can be initiated in several ways under U.S. patent law:

- If an examiner discovers that two or more pending applications have substantially the same claims and it has been found that these claims are indeed patentable, an interference will be "declared".
- If two or more applications have similar or overlapping claims, one or more claims ("counts") to the common subject matter of the application will be proposed, and the applicants will be given the

1997–4/91/0077/$06.00/1
© 1991 American Chemical Society

opportunity to adopt the proposed claims for interference purposes.

- If a patent inadvertently issues without the discovery of conflicting claims in another pending application, perhaps because the applications were assigned to different examining groups, the second applicant can request an interference.

- If the owner of a pending application recognizes that a patent that has issued claims subject matter disclosed but not claimed in his or her application, the claims of the issued patent can be added ("copied") to the application or a continuation application with a request that an interference be declared. A continuation application is one that has the same specification and the same or amended claims as in the first-filed application. This copying must be done within a year of issuance of the patent.

- Finally, if two conflicting patents issue, there can be a lawsuit between the parties to determine priority. In some cases patentees have succeeded in filing an application for reissue of their patent, copying the claims of the other patent, with a request for an interference proceeding. Except in court actions, the matter goes to the Board of Patent Interferences to determine which inventor has a right to the patent.

The applicant with the earliest effective filing date, which may be that of an earlier application, is designated the "senior party", and the later applicant, or applicants if there are more than two in the interference, is called the "junior party". It is a substantial advantage to be the senior party because that applicant can simply rely on his or her filing date without further proof. The burden of proof that the invention was made earlier is on the junior party, with all the perils of a vigorous challenge to the sufficiency of his or her evidence. For this reason, if no other, it is extremely important that the preparation and filing of the patent application be done as promptly as possible after an invention is recognized.

Procedures Encountered in an Interference

An interference is a complex legal proceeding, hedged in with rules and deadlines—71 of the some 250 Rules of Practice from the Code of Federal Regulations relate to interferences. The conduct of the procedure is of course in the hands of the attorney, who may well call in the assistance of a specialist in interference practice. The principal steps are a preliminary statement by each party, indicating the dates and evidence on which their positions will rely; a motion period, during which there may be challenges to the proposed counts and the rights of each other to make the claims; and a testimony period, in which witnesses may be examined and cross-examined on details of the evidence relating to their knowledge and actions. The junior party takes testimony first, and the senior party has the choice of relying on his or her filing date without taking testimony if that appears to be advantageous. The chemist is involved in this process in two ways: (1) to produce the written records of experiments, and (2) to give oral testimony if needed.

Records

Notebook records are vitally important in an interference proceeding. Every working chemist writes hundreds or thousands of pages of notebook records, and obviously the vast majority of these will never be needed in a patent case. It is easy to slip into careless habits in keeping notebooks, but on the rare occasions when the notebook must be produced, it is absolutely necessary that it be a record that is sufficiently complete that another chemist can understand and reproduce the work and that there is a witness who can give corroborating testimony if needed.

These are the important points of a good record:

1. It should be in a bound notebook. Loose records are easily challenged and hard to support because the dating cannot be tied to other contemporary records.

2. Experiments should be recorded in chronological order. Skipped or blank pages or pages dated out of order create a suspicion of tampering with the record.

3. Each experiment should be dated when it is started, and if the work carries over more than 1 day, each day's entry should be dated.

4. The experiment should start with a clear statement of the objective.

5. All essential facts should be recorded, such as equipment used, conditions, times, materials used including source and quality, yields, characterizing data, and so on. Abbreviations and codes should be chosen and used in an unambiguous way.

6. The record of an experiment that takes more than one page should make definite references to previous and following pages so it can be followed, for example, "continued on p.—", "continued from p.—".

7. If a standard or routine procedure is being followed, a reference to the location of a full description should be made.

8. The record should draw a conclusion if possible. A conclusion may not be needed if the experiment is one of a routine series, but experiments that explore new conditions or are aimed at making a new composition should conclude with an evaluation of the results. This step is important because recognition of success is an important element in the reduction to practice of an invention. Unnecessary derogatory comments about the results should not be made—the results may be valuable in a different way from what was anticipated when the experiment was started.

9. Analytical or other test results should be attached or copied into the record, or if they are too bulky, reference should be made to where they can be found.

10. Any unused portion of a page should be struck out to forestall any challenge that the record has been augmented at a later date.

11. Entries should be in permanent ink.

When the record is complete, or when there will be some delay before the work will proceed, the notebook page should be signed and witnessed promptly. These signatures must be dated. The purpose of witnessing is to provide corroboration of the existence of the record at the date of signing by a person who can testify later if needed. The reason is that an inventor's unsupported testimony on his or her own behalf is considered under the law to be self-serving. Many interferences have been lost because no corroboration was available for the inventor's testimony. Witnessing should be done no more than a few days after entries are made. Witnessing that is unduly delayed is little better than no witnessing at all. Preferably the witness should be someone who has observed and understood the experiments—the laboratory technician may be a good witness—but in any event the witness should have read and understood the entries and should be a person who can reasonably be expected to be available for several years after the date of signing. The witness should not be a potential co-inventor for the reason already mentioned.

Consistency in the keeping of laboratory notebooks is extremely important. Occasionally in patent litigation a judge will accept the accuracy of notebook records despite a lack of proper witnessing if they have been kept chronologically, in a bound notebook, and according to a well-established pattern. However, a properly witnessed record is much safer.

Testimony in Interference Proceedings

An important interference proceeding will eventually reach the period for the taking of testimony. In the usual procedure the attorneys for both parties meet in an attorney's office or a conference room in the presence of an officer qualified to take sworn testimony, such as a notary public or a court clerk. A public stenographer records all that is said, or a recording is made of the testimony, as in a court trial. Witnesses are called, generally one at a time, and are questioned (examined) under oath by counsel for the party who calls them. They are then cross-examined by counsel for the opposing party. The transcripts of these proceedings (depositions) are then submitted to the Board of Patent Interferences of the U.S. Patent and Trademark Office (PTO) as part of the record that serves as the basis of their decision as to which inventor is entitled to receive the patent. Sometimes the decision is that each party should receive a patent to different embodiments of the invention.

Witnesses who may be called include the inventor, the person who witnessed the notebook record, others who have knowledge about how the invention was made, other chemists who may have confirmed the invention by repeating the experiment, and chemists who provided analytical data or test results for the inventor. The testimony must be candid. Any attempts to obfuscate the facts will surely be caught by opposing counsel and will throw doubt on the entire testimony. Witnesses should answer questions directly, with no attempt to volunteer extraneous opinions. The attorneys will usually meet with witnesses beforehand to brief them on what to expect. Tactics in the questioning must be left to the attorneys, although they will appreciate suggestions or background information about the opposition's interests and prior activities that may provide clues on how to approach opposition witnesses. The exposure to legal procedures that goes with the taking of testimony is an unusual but interesting experience for technical people. Sometimes the parties agree to submit sworn statements of the witnesses instead of depositions. These declarations or affidavits are sworn and must also be accurate and truthful.

Discovery

Another aspect of interferences as well as other patent litigation that can touch the activities of chemists and engineers is the legal process called "discovery". A party to an interference may suspect that the opponent has records that might help the case. A court or board order can be obtained requiring production of all records and files related to the counts of the interference, which would include notebooks, memoranda, correspondence, and reports. The technical person should keep this possibility in mind when recording findings and conclusions. In general, technical people should make their records strictly factual and avoid any attempt to make quasi-legal judgments about such matters as patentability or patent infringement. Unnecessary written conclusions or admissions, which may well turn out to be incorrect in the light of legal analysis or further experimentation, could jeopardize a case if they have to be produced under a discovery order.

Settlements

Interferences that involve inventions considered to be valuable by the parties can be hotly contested and extraordinarily costly. Probably the most extreme case involves the invention of isotactic polypropylene, on which patent applications were filed by five parties in 1954–1956. U.S. Patent 3,715,344 was granted in 1973 after many years of arguments, counterarguments, taking of depositions, and complex legal maneuvering, both in the PTO and in the federal courts. Litigation challenging this patent was not concluded until 1985. Polypropylene is a product being made commercially by several companies at the rate of more than 2 billion pounds per year; such production means that the potential royalties to the winner of an unchallenged patent are enormous.

Few interferences involve inventions as valuable as polypropylene. However, it is relatively common for the parties to decide that the potential profit from their inventions cannot justify an expensive legal proceeding. In this circumstance it is common for the parties to arrange a "settlement", in which both parties agree to have their attorneys compare their invention records and decide which one is the first inventor under the patent law. The settlement contract normally includes an agreement that the loser is entitled to a license to the patent obtained by the winner, which may be royalty-free or royalty-bearing. The loser then concedes priority and withdraws that patent application. A copy of the settlement agreement must be filed with the PTO.

Proof of Priority; Diligence

Barring settlement, the Board of Patent Interferences of the PTO ultimately decides which applicant is entitled to a patent. The two parts in the making of an invention are (1) the concept and (2) reduction to practice. The concept is the mental formulation, or the idea, in sufficient detail to provide the basic elements of the invention. Reduction to practice is the actual physical accomplishment of the invention. An invention is not complete until reduction to practice has been carried out, and the decision on who is the first inventor hinges primarily on which applicant was the first to conceive and to proceed without delay to reduction to practice.

A second way of completing an invention is simply to file a patent application without benefit of an actual physical demonstration. This method is called "constructive reduction to practice". Sometimes this is the only practical procedure, as in a conception that would involve massive, prohibitively costly, or remotely located projects. This course obviously requires an unusual degree of understanding and technical knowledge by the inventor to provide a disclosure that can withstand a challenge to its operability in the case of later litigation.

Exceptions to the reduction to practice basis for establishing priority occur if the junior party (last to file) can show the earliest date of conception and reasonable diligence in reducing the invention to practice, even if the date is later than the other party's; or the junior party can win by showing that even though his or her date of conception was later than that of the senior party, reduction to practice was completed before filing by the senior party and that the senior party was not diligent in filing an application. Diligence is a somewhat subjective matter, and the courts examine the circumstances closely in each case in an effort to draw fair conclusions about whether diligence has been proved. The most that can be said for certain is that a technical finding that appears to be important should be given prompt, well-documented, and continuing attention by all persons concerned, especially including the inventor and the attorney who files the application. The interest of the patent system is in promoting prompt disclosure of new findings to the public, and unnecessary delays in moving toward that goal are frowned on by the courts. The party who promptly files a patent application will almost certainly win over one who is dilatory. An applicant who can be shown to have suppressed, concealed, or abandoned the invention has no standing and cannot be granted a valid patent under any circumstances.

The Decision; Issuance of a Patent

Decisions of the Board of Patent Interferences can be appealed to the Court of Appeals of the Federal Circuit or to the appropriate U.S. district court. When an interference is settled either by the Board or by agreement between the parties, or after appeal, the case goes back to the primary examiner, who then issues a Notice of Allowance if no further information that might throw doubt on the patentability of the claims has come to light meanwhile. Issuance of the patent follows in due course, as described in Chapter 7. Only about 1% of all U.S. patent applications become involved in interferences, but a heavy burden of legal costs goes into their resolution. Prompt consideration should be given to patent action when an invention is recognized, especially in a competitive field, to avoid getting into an interference situation.

9

Patent Infringement
and
Patent Claims

The claims are the heart of a patent. They are the legal description of the grant of rights to the composition, process, or apparatus that the owner may exclude others from using for the life of the patent. Claims are like property—they can be bought, sold, rented (licensed), or allowed to lie fallow. They are subject to trespass, which in patent terminology is called infringement. A patent owner protects claims against infringement either by persuasion or by suit, normally in a federal district court.

Patent infringement suits are complex and expensive, often costing the parties hundreds of thousands of dollars and, in more recent cases, millions of dollars. For this reason as well as their traditional high ethical standards, companies in the chemical industry generally make an effort to avoid infringement of the patents of others. It is in the best interests of all to respect valid patents, and it is of course foolish to embark on a program of research and development that leads to a product or process that cannot profitably be exploited because of the patent situation.

Nature and Kinds of Infringement

The three types of infringement are as follows:

1. Direct infringement occurs if a patented composition or apparatus is made, used, or sold or if a patented process or product is used.
2. Contributory infringement occurs if a material, device, or a part of a device is sold specifically for use in a patented composition, process, or completed device. Such sale enables a customer to engage in direct infringement. Legal liability is found with both parties, and, in theory, equally with both parties. However, the

1997–4/91/0085/$06.00/1
© 1991 American Chemical Society

contributory infringer usually bears the principal legal burden. Infringement does not necessarily occur from a sale of a material that has other substantial uses unless a definite recommendation to engage in an infringing use can be shown on the part of the seller.

3. Induced infringement results if a customer is encouraged to use methods or products that are claimed by patents of others.

Infringement can arise from the introduction of a new process or product or from a change in an existing process or product in the face of existing patents. It can also arise from newly issuing patents claiming some aspect of an established process or product. Constant vigilance is necessary to guard against incurring liability for patent infringement in commercial operations. The patentee, assignee, or licensee is individually responsible for policing the use by others (infringement) of the subject matter of the patent; no governmental agency automatically performs this function. Whenever a change is made in product or a manufacturing process or a new product is introduced, it is advisable to consult with an attorney or agent to ensure freedom to practice.

Watching for Infringement; Reading Claims

Responsibility for awareness of patent claims that may be a problem must rest to a large extent on the chemists and engineers who are engaged in studying new technology. They should have a reasonable understanding of how to read patent claims so they can call attention to potential problems and plan their studies to avoid future conflicts. This chapter tries to provide guidance on the basics of claim interpretation.

The Real Estate Analogy

A patent claim can be likened to the description of a parcel of land in a real estate deed. The limits of the claim serve the same function as the metes and bounds that locate and define the perimeter of a building lot. The owner of the land has the right to exclude all others from trespassing, for example, by building a fence around the property. A gate can be put in the fence and a fee charged for entering, just as a patent owner can charge a royalty for use of a claim. Patent property differs from real property in the sense that others can obtain claims to improvements on the broad claim, as though another person could put a house on the unimproved property of the first owner. If this could happen in the real estate case, the landowner could prevent access to the property by the house owner, and the house owner could lock the door to the landowner. On a cold or rainy day, they would likely reach an agreement to give each other access to their respective properties. In patent jargon, this agreement produces cross-licensing of the patents held by each for the mutual benefit of each party.

Dominating Claims

In the patent sense, the broad claim is said to "dominate" the claim to an improvement, and patent owners who find themselves in such a relationship, assuming that the improvement is the commercially most desirable product or process, usually arrange a cross-licensing agreement so that both patent owners can take advantage of the inventions.

A simple case in patent claiming would be the relationship of a hypothetical claim reading, "A bladed instrument for driving screws"—that is, a screwdriver—and the device claimed in another, actual patent to, "A cross-bladed instrument for driving screws"—that is—a Phillips screwdriver. If the first claim had been in force when Phillips made his invention, he would have needed a license to make his improved screwdriver, and the owner of the dominating claim to any bladed instrument would have needed a license from Phillips in order to share in what became a large market for cross-bladed screwdrivers.

Consider the following claims from two patents in the chemical field. U.S. Patent 3,033,835 claims:

> A sulfur-vulcanizable, elastomeric copolymer of at least two straight-chain alpha-olefins of from 2 to 10 carbon atoms and an ethylenically unsaturated bridged-ring hydrocarbon containing at least two ethylenic double bonds, at least one of said double bonds being in one of the rings of the bridged ring, present in the copolymer in an amount imparting sulfur-vulcanizability, said hydrocarbon having from 7 to 20 carbon atoms, the total straight chain alpha-olefin content of said copolymer being at least 50%.

U.S. Patent 3,093,620 claims:

> A sulfur-curable copolymer of ethylene, at least 1 alpha-olefin having the structure $R-CH=CH_2$ wherein R is a C_1-C_3 alkyl radical, and 5-alkenyl-2-norbornene, said copolymer containing at least about 20% ethylene units by weight, at least 25% of said alpha-olefin units by weight, and at least about 0.03 gram-

moles per 100 grams of said copolymer and not over 20% by weight of said copolymer of said 5-alkenyl-2-norbornene.

From a purely technical standpoint, the first claim dominates the second. Ethylene and an alpha-olefin (R–CH=CH$_2$, R being C$_1$–C$_3$), are "at least two straight-chain alpha-olefins of from 2 to 10 carbon atoms"; 5-alkenyl-2-norbornene is an ethylenically unsaturated bridged-ring hydrocarbon as described; the ethylene and alpha-olefin content are within the broader limits of the first claim, and the defined concentration of 5-alkenyl-2-norbornene is in fact sufficient to impart sulfur vulcanizability to the polymer. Therefore the claim of U.S. Patent 3,033,835 includes the scope of U.S. Patent 3,093,620. Assuming that the patents are valid, the owner of the latter would need a license from the owner of the former to make, use, and sell the olefin copolymers that include the alkenylnorbornene. The owner of U.S. Patent 3,033,835 has the rights to such polymers containing bridged-ring diolefins broadly but would need a license from the owner of U.S. Patent 3,093,620 if the alkenyl norbornene polymers were the preferred commercial product.

To infringe a patent claim, every critical element of the claim must be met. A composition containing A + B + C + D would infringe a claim to A + B + C, but a composition containing only A + B would not. A patent claim should be read one phrase at a time, and the reader should ask after each phrase whether a proposed, potentially infringing, product or process would include that component or step or the equivalent component or step, as the case may be. If the answer after any phrase is an unequivocal "no", any claim of infringement can probably be safely dismissed.

Narrow and Broad Claims

Consider the following claim from U.S. Patent 2,892,722.

A method of preserving celery in a firm and solid condition for an extended period of time comprising cutting off the ends and leaves of the celery, washing the celery in cold water, cutting the stalks of the celery into lengths from three to four inches, providing a solution comprising approximately the proportions of four ounces of salt, minor portions of garlic, mustard, coriander, allspice, ginger, cinnamon, chiles, cloves, black pepper, bay leaves, white pepper, and mace, providing approximately three hot peppers, approximately ten ounces of dry wine vinegar, and approximately three quarts of water, placing the lengths of celery in the said solution, and allowing said celery pieces to stand in said solution for more than eight weeks.

Although the terminology may seem peculiar at first glance, claims with as many limitations as this one are characterized as being "narrow". Applying the real estate analogy, this claim can be thought of as a long, narrow lane hemmed in by many fence posts. The invention is no doubt novel and useful, but it could likely be avoided by working outside the defined limits or by omitting one or more of the many required ingredients with little change in its usefulness.

By the way of contrast, the first claim of U.S. Patent 3,051,677 would be described as "broad" because it has so few limitations:

1. An elastic copolymer of about 70% to 30% by weight of vinylidene fluoride and from about 30% to 70% by weight of hexafluoropropene.

The entire range of copolymers of vinylidene fluoride and hexafluoropropene that have elastomeric properties is included. This claim would be difficult to avoid.

Independent and Dependent Claims

Most patents have one or two "independent" claims, which recite the critical limitations of the invention in their broadest ranges. These are followed by a series of "dependent" claims, each of which incorporates by reference the features of the independent, "main" claim, or other dependent claims, with the addition of particular specific compounds or conditions within the scope of the main claim. The purpose is to include claims of increasingly narrow scope to fall back on in case the main claim should be found invalid in the face of later discovered prior art.

"Jepson" Claims

Another form of claim often used when the invention is an improvement on existing technology, known in U.S. practice as a "Jepson" claim, recites in a preamble the essential features of the established art, for example, a process for manufacture of a known product. All that appears before the linking phrase "the improvement wherein" (or similar language) is old in the art, and the language that follows the linking phrase is what is in fact the new invention. German claims are normally in this form, the linking phrase being "characterized in that". Many persons not familiar with patent-claiming practices react with dismay at their first reading of such a claim—"How can they patent what everybody already knows?" The preamble is only a recitation of "what everybody knows" to set the framework for claiming the inventive improvement. The entire claim must be read carefully to determine what is actually new. The patent would not have been granted unless the examiner were satisfied that a patentable invention had been made.

Claims of Foreign Patents

The chemist should also realize in studying foreign patents that several countries publish patents without examination, leaving interpretation of their scope to the courts. Also, the European Patent Office, West Germany, the Netherlands, and Japan automatically publish unexamined applications under their deferred examination systems 18 months after the first, or priority, filing date. The claims of such publications have no force although the publication is, of course, part of the prior art (as a publication) against any later patent application on the same invention. The claims should be considered only to be a declaration of what the applicant hopes to have allowed. The finally granted patent, if any, may have claims of much more limited scope.

Pitfalls in Reading Claims

The chemist or engineer considering the infringement question must be cautious in reading and interpreting patent claims, because there are many pitfalls. The first claim of a patent is most often the broadest, but not always. All the claims need to be scanned, because claim 13 of 20 may be the one that creates a problem.

At this stage the claims should not be read too literally. For example, it cannot be assumed that the use of compound Q in place of compound P in a composition avoids infringement. If Q serves the same function as P and in the same way, a court may apply the "doctrine of equivalents" to conclude that the claim is being infringed. Also, it cannot be assumed that using 45% of compound M avoids a claim to a composition containing 50–70% of M. These are questions that require detailed consideration by patent experts.

A detailed explanation of the doctrine of equivalents in U.S. patent law is beyond the scope of this text. The now classic U.S. Supreme Court decision of Graver Tank and Manufacturing Co. v. Linde Air Products Co., 339 U.S. 605 (1950), Vol. 85 *U.S. Patent Quarterly* page 328, is a decision regarding the substitution of one chemical for another in a soldering flux. A more recent court interpretation affirming the doctrine of equivalents is found in the Court of Appeals of the Federal Circuit case of Hybritech, Inc. v. Monoclonal Antibodies, Inc., 802 Fed. 1367, Vol. 231 *U.S. Patent Quarterly* page 81 (Court of Appeals of the Federal Circuit, 1986).

The opposite view to the doctrine of equivalents is often referred to as the "doctrine of a different animal", and is found in the U.S. Supreme Court decision in Westinghouse v. Boyden Power-Brake Co., Vol. 170 *U.S. Patent Quarterly* page 537 (1898). The holding is that even though the words of a patent claim do in fact clearly describe an accused subject matter, still the invention of the patent and the accused subject matter are so completely different—really entirely different animals—that infringement will not be found.

What To Do When There May Be Infringement

Whenever a patent claim is encountered that raises even a remote question of possible infringement, the technical person must seek the advice of a patent professional. If the organization includes patent liaison people, they may be able to dismiss the patent from consideration on technical grounds because of their greater familiarity with patent practice and the prior art. If there are questions of legal interpretation, the matter should go to an attorney, with a detailed description of what product or process or recommendation to a customer is being contemplated and any prior art that should be considered by the attorney in interpreting the patent.

Legal Analysis of the Claims

The attorney will most likely obtain from the U.S. Patent and Trademark Office (PTO) a copy of the "file history" of the patent, which includes all the communications between the patent owner and the examiner during prosecution of the application. The attorney will consider such questions as whether there is pertinent prior art that the examiner did not consider, whether the examiner interpreted the art properly, the extent to which the doctrine of equivalents may apply, how restrictively quantitative limits of the claims should be interpreted, and so on. For instance, with respect to the question of whether using 45% of compound M would infringe a claim to 50–70% of M, the file history might show that the applicant originally tried to claim 35–90% of M but had to retreat to the narrower range either because of art cited by the examiner or because the experimental data did not justify the broader range. In such a case, the attorney might conclude that the claim would probably be restricted by a court to the 50–70% range. If no such situation exists, the attorney might conclude that use of 45% M would probably be held to be an inconsequential difference (a "colorable deviation") and would infringe the claim. Limitations imposed as a result of changes in scope accepted by the applicant during the prosecution are said to result in "file wrapper estoppel", meaning that the applicant is barred—that is, "estopped"—from extending the claim beyond its literal language if during prosecution it was required that the claim be narrowed to secure allowance.

The attorney will have previously made a judgment as to whether the proposed course of action would in fact infringe the claim if valid. On the basis of the considerations just discussed, he or she will then formulate an opinion on the probable validity of the patent in question. If the opinion is that the patent would be infringed but is not valid, a business decision must be made on whether to proceed in the face of the patent. This decision requires consideration of the liability that would be incurred if litigation should follow and for some unforeseen reason should result in an unfavorable finding by the courts. Potential profitability of the proposed product or process must be taken into account, as well as the substantial cost of a lawsuit. It is in the best interest of the patent system that invalid patents should not be allowed to stand in the way of the use of technology, and an affirmative decision to proceed should be made, barring business circumstances that stand in the way.

The Options When Infringement Is Found

If the attorney concludes that the patent would be infringed by the proposed product, process, or apparatus and that it appears to be valid, the first alternative is to change the product or process to avoid infringement if possible. This course of action is perfectly legitimate. The purpose of the patent system is to encourage continuous improvement of technology, and it is frequently possible to "invent around" all but the broadest, most basic patent claims. The second alternative is to negotiate with the patent owner for a license if that course is economically acceptable.

There are no established standards for royalties. They may range from a fraction of a percent of the sales price for very large-volume products such as the major plastic resins to 25% or more for small-volume, high-value products such as potent pharmaceuticals. The third option is to file a petition for reexamination of the patent with the PTO. By the petition, the attorney would be attempting to have the patent declared invalid and thus render moot the claim of infringement.

＊　　＊　　＊

The other side of the infringement question is the problem of exploiting patent rights that the applicant owns. This is the subject of Chapter 10.

10

Making Use of Patents: Enforcement

A patent can be an extraordinarily valuable property. Personal fortunes and giant industries have been built on patent rights. Hall's aluminum recovery process (U.S. Patent 400,766), Baekeland's thermosetting resins (U.S. Patent 942,699), Carothers' nylon (U.S. Patents 2,130,947 and 2,130,948), Carlson's xerography (U.S. Patent 2,287,691), Land's instant photography (U.S. Patent 2,543,181), Teal's silicon transistors, Hall's synthetic diamonds, and Noyce's integrated circuit (U.S. Patent 2,981,877) are only a few of the best-known examples of inventions that resulted in major industrial developments as well as personal fame and reward. In early 1977 the Bausch and Lomb Company paid $14 million to settle a patent suit on hydrogel soft contact lenses. In 1989, Keebler, Nabisco, and Frito-lay paid Proctor & Gamble $125 million to settle a patent suit on cookies. Of course, the great majority of patents have limited value, but the stakes are huge for the winners, and it is seldom possible to recognize a winner in its early stages.

Patents can also be a liability. Attempts to enforce poorly drafted or carelessly prosecuted patents can backfire. Recently a company sued alleged infringers of a patent that the court ultimately characterized as lacking novelty and utility, as being obvious, as having an inadequate disclosure and ambiguous claims, and as being invalid in any case because of use and sale more than a year before the application was filed. The court held that the charge of infringement was made in bad faith because the patent owners knew or should have known that the patent was unenforceable, and the court ordered that the owners pay $237,000 for the defendants' attorneys' fees—an expensive lesson for those who misuse the patent system.

1997–4/91/0095/$06.00/1

Enforcement of Patent Rights

A U.S. patent is in force as of noon on the day it issues. Until then there are no enforceable rights, although at times patent applicants start licensing discussions or warn potential infringers on the basis of an allowed application. Such discussions are the applicant's choice, because the U.S. Patent and Trademark Office (PTO) will release no information about a pending application.

If an application is pending and it appears that a competitor is using or preparing to use the invention, the applicant can file a petition in the PTO to make the application "special". This step will accelerate consideration by the examiner and cause earlier issuance of the patent if it is allowed. Also, an application for reissue of an existing patent can be filed if its claims do not cover the competitor's use adequately but the disclosure actually provides basis for claims that would do so.

The right to obtain damages for infringement does not begin until the alleged infringer is put on notice by the patent owner. This notice can be done by marking a patented product that is for sale with the number of the patent, by direct communication through a letter, or by actually filing suit in one of the federal district courts. Most companies in the chemical industry respect valid patent rights, and unless there are substantial reasons for doubting the enforceability of a patent, some accommodation is usually reached after an initial communication from the patent owner. Failing this, the alternative is to file an infringement suit.

An infringement suit is unwise unless the patent owner is quite certain that the patent is enforceable. Patents may be unenforceable for many reasons. In some cases, the claims can be avoided by deletion of required components, as appears to be the case with the patent on a method for pickling celery discussed in Chapter 9.

Patents may be impractical to enforce because of the nature of the business to which they pertain. U.S. Patent 3,925,873, whose cover is shown in Figure 10-1, seems to present this sort of problem. The claims, as implied by the abstract, are to a method and apparatus for aligning the backstays of a bicycle. The apparatus consists of a simple supporting device, a notched stick, and a long rubber band. The method consists of bending the backstays with the stick until they are equidistant from the lines defined by the rubber band. Bicycle repair is not an organized industry but is done in thousands of small, single-owner shops scattered throughout the country. This patent appears to claim a novel and useful invention, but it is difficult to visualize any practical way that it could be

United States Patent [19]

Mecum

[11] **3,925,873**

[45] **Dec. 16, 1975**

[54] **BICYCLE FRAME ALIGNMENT APPARATUS**

[76] Inventor: **Robert C. Mecum**, 5728 Monona Drive, Madison, Wis. 53716

[22] Filed: **Feb. 3, 1975**

[21] Appl. No.: **546,193**

[52] **U.S. Cl.** **29/271**
[51] **Int. Cl.²** **B25B 27/14**
[58] **Field of Search** .... 29/271, 407; 269/54.4; 248/298, 176; 33/193, 180 R; 72/388, 705, 458

[56] **References Cited**

UNITED STATES PATENTS

2,479,723	8/1949	Brown	33/193
2,590,722	3/1952	Otis	33/193
2,667,798	2/1954	Beasley	72/458

FOREIGN PATENTS OR APPLICATIONS

135,655	5/1952	Sweden	248/298

Primary Examiner—James L. Jones, Jr.
Attorney, Agent, or Firm—Theodore J. Long; John M. Winter; Harry C. Engstrom

[57] **ABSTRACT**

Apparatus and method for aligning the rear stays of a bicycle frame. The apparatus has an elongate bar slidably engagable within an elongate tube. Seat and head tube spindles secured in upright positions at opposite ends of the elongate tube and bar are inserted into the head and seat tubes of a bicycle frame. An elastic band is attached to the rear dropouts and stretched around the head tube of the bicycle frame. A stay aligning tool is placed against either side of the seat tube and levered against the rear stays of the bicycle frame until the distance from the elastic band to the seat tube is equal on each side while maintaining the rear hub width between the dropouts.

2 Claims, 3 Drawing Figures

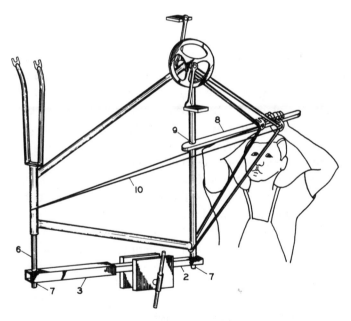

Figure 10-1. A patent that would be difficult to enforce.

widely enforced, because each infringer would have to be sued individually. This and the celery-pickling patent are extreme cases, but the questions about enforceability that they represent should be considered before a patent is applied for and must be considered before enforcement is attempted.

Enforceability Studies: Disclaimers

Patent owners should ask an attorney or legal firm to make an enforceability study before any action is started. This study can involve an even more extensive search of the prior art than was made in preparing the patent application, a review of the file history as in an infringement study (*see* Chapter 9), interviews with the inventor and other technical people who worked on the development, and a detailed study of the disclosure and working examples to be sure that there are no serious errors that could throw doubt on the operability of the teachings of the patent.

It may be recommended that any weak claims of the patent should be disclaimed before starting an infringement action to reduce any prejudice that their presence might induce in a judge's view of the case. Most issues of the *Official Gazette* include a record of such disclaimers.

Suing for Infringement

Infringement suits are filed in federal district courts. Interrogatories—that is, questions about the alleged infringing activity—are submitted to the defendant. Interrogatories may also be submitted by the defendant to be answered by the plaintiff. These may or may not be allowed by the judge. Discovery (deposition testimony, document production, etc.) is usually taken by both parties. This discovery is similar to that discussed in the interference section. Before the trial, briefs are exchanged, as are lists of witnesses, exhibits, and issues of fact and law to be proven. To establish the case, evidence of ownership of the patent and of apparent infringing conduct by the defendant must first be introduced.

At trial, expert witnesses may be called by both sides to give technical testimony about the meaning of the patent claims and the nature of the accused actions of the defendant. There will be arguments about whether differences in materials or process steps used by the alleged infringer are functionally equivalent to those claimed in the patent and whether differences may have been selected to avoid literal infringement but are only "colorable deviations" from the limits of the claims. Discovery orders may be issued to ascertain facts and to obtain records relative to the case. Few judges are technically trained, and a master, usually a professor experienced in the scientific discipline of the patent, may be appointed by the judge to study the matter and make factual recommendations. Oral arguments will be made and witnesses called, depending on the judgment of the attorneys handling the case.

Defenses in Infringement Suits

The defenses to be expected on the part of the alleged infringer include

1. The patent claims were issued improperly to nonstatutory subject matter (Title 35 of the U.S. Code, Section 101 [35 U.S.C. 101]).

2. The claim improperly or indistinctly described the invention or did not enable one of ordinary skill in the art to perform the invention (35 U.S.C. 112).

3. The inventor failed to describe the best mode (35 U.S.C. 112).

4. There is no infringement because not all the critical limitations of the claims are being used, or some distinct and nonequivalent element or step is being used rather than what is claimed, or the accused operation is outside of some critical range of the claimed composition or process step.

5. The patent is invalid because it lacks novelty or is obvious in the light of the prior art or because it fails to provide an adequate disclosure of how to make or use the invention (35 U.S.C. 102 or 35 U.S.C. 103).

6. The patent was obtained as a result of fraud on the PTO, which could include inaccurate or incomplete statements in the oath with respect to inventorship or prior use or sale of the invention, withholding from the examiner or misrepresentation of knowledge about the prior art, or misrepresentation of facts during prosecution, as in test results reported in an affidavit or an example (Title 37 of the Code of Federal Regulations, Section 1.56 [37 C.F.R. 1.56] and/or 37 C.F.R. 1.97).

7. The complainant lacks the right to sue for infringement because the alleged infringer was allowed to proceed unchallenged for an extended period before filing suit—a doctrine known in law as "laches", or undue delay in asserting a legal right.

8. The patent owner has misused the patent by such actions as mismarking products with patent numbers that do not apply, by trying to force the purchase of unpatented products as a condition of a patent license, or by other actions that violate the antitrust laws.

Declaratory Judgment Actions

The filing of an infringement suit allows the alleged infringer to file a counteraction asking that the patent be declared invalid for one or more of the reasons mentioned. Even the threat of an infringement suit allows the filing of a declaratory judgment suit challenging the patent. These responses can be expected in an important case. The choice of the district in which the suit is filed is up to the complainant, limited by certain rules about where the parties do business and where the alleged infringement is taking place. A jury trial can be requested rather than trying the case before a judge. Indeed, jury trials are more often requested by defendants accused of infringing a patent, especially in cases where the defendant's chances of proving noninfringement are not great. In such circumstances, the defendant may take the chance that the jury will be confused by the complexity of the patent and technical issues and thus render a finding in favor of the defendant.

Injunctions and Penalties

If the complainant succeeds in proving infringement, the court can issue an injunction against further use of the invention and assess damages for the infringement that has taken place. If the infringement is judged to be willful because the infringers knew and understood what they were doing, the patent owner's attorney's fees may also be assessed. This assessment is not done when the infringers can show that they had a reasonable doubt that the accused activities constituted infringement or that the patent was valid.

Ill-considered litigation on an invalid or misused patent can yield heavy penalties for a patent owner, just as willful infringement of a valid patent does for the infringer. The patent owner can lose on any of the grounds for defense already mentioned. It may be found that there is no infringement or that the patent is invalid. Sometimes the decision is that the patent is invalid but would be infringed if it were valid. It is also possible for a patent to be held valid and infringed but unenforceable because of inequitable conduct on the part of the inventor, attorney, or agent. The reason for such a decision is to provide a complete analysis for the appeal court in case that court should disagree on the validity question.

A distinct hazard for those who sue and lose on grounds of fraud or antitrust violations is the assessment of the defendant's legal costs and possibly triple damages. Fraud or misuse of patents can be an expensive mistake. Complete candor in dealing with the PTO and scrupulous adherence to legally allowable uses of patents will avoid such problems.

Infringement by the Government

A special case of enforcing patent rights is presented when the federal government has made unauthorized use of a patented invention. Suit can be filed in the Court of Claims. The government is the defendant in this case. The procedures are essentially the same as in a conventional infringement suit, and damages may be recovered by patent owners in this way.

Working Requirements in Foreign Countries

Successful prosecution of an infringement action on a valid patent will preserve the patent owner's right to maintain the legal 17-year monopoly. The patent owner may choose to license the patent to the loser and perhaps to others, or, in the United States, not to use the invention at all. In contrast to U.S. law, many foreign countries have "working" requirements. Actions that will satisfy the working requirement vary widely. In certain countries the act of simply publicizing the patent is sufficient, in others licenses must be offered, and in some the invention must actually be practiced by the patent owner or a licensee. Otherwise the patent will lapse. Patent attorneys deal with this problem through their foreign associates.

Importation

U.S. patent law now forbids importation of products made by a process that is the subject of a U.S. patent claim. Occasionally the owner of a patent to a process that produces a product that does not itself have patent protection recognizes that competitors are importing articles made by the patented process. This problem can be handled by an appeal to the U.S. Customs Service. If it can be established that the imported product is in fact being made by a process on which there is a U.S. patent in force, an order will be issued to halt further importation.

If an article that is patented in the United States is fabricated outside the border and imported, legal actions that are fairly speedy (that is, usually completed and decided within 1 year) are available under Section

337 of the International Trade Commission statutes. A complete discussion is beyond the scope of this text. Local patent counsel should be sought, because the impact and scope of the law is being modified every year.

Licensing

Considerable licensing of patent rights has often occurred in the chemical industry, and as the cost of research and development continues to increase, chemical companies frequently seek rights to technology developed by others rather than duplicate technical effort. The negotiation of patent licenses is a specialized business problem that is beyond the scope of this discussion. Several of the books mentioned in the Bibliography offer advice on this subject. The publications of the Licensing Executives Society are a continuing source of information in this field.

Royalties paid for patent rights vary widely, depending on the volume of business anticipated, the nature of the particular business, and the strength of the patent. Sometimes a single payment for paid-up rights for the life of a patent will be negotiated. Running royalties can vary from a fraction of a percent of selling price for a large-volume material to 25% or more for high-value products. Sliding scale royalties depending on volume are common.

Licenses can be exclusive or nonexclusive. Most licenses provide for adjustment of royalties if more favorable terms are offered to any other licensee. There is no legal requirement that licenses be offered to all who want them unless a court finds clearly demonstrated discrimination against a class of users who are put at an unfair disadvantage because they were denied a license.

The right to practice a patent claiming the use of a product may be conveyed by what is called a "label license". The label applied to the product states that the price includes the right to practice the claims of one or more named patents, and a label license usually indicates that a separate license can be obtained from the patent owner at the same royalty rate if the user wants to obtain the product from another source. Quite often companies will take licenses to patents on uses of products they want to offer so they can extend to their customers the right to practice the invention.

Care must be taken in negotiating patent licenses to avoid arrangements that extend the period of royalty payments beyond the life of the patent or that violate the antitrust laws by requiring a "tie-out", an agreement not to sell a competitor's product, or a "tie-in", an agreement to buy other unpatented products as a condition of the license. Field-of-use restrictions are generally acceptable. For instance, one licensee may be given the right to use a patented compound as a rubber chemical and another to use it as an herbicide.

At one time it was common for patent licenses to include an agreement that the validity of the patent could not be challenged by the licensee. Court decisions have made such agreements unacceptable, and declaratory judgment suits by licensees challenging the validity of patents to which they are licensed are frequently seen.

Patent licenses are a substantial source of income for many individuals and corporations. One of the petrochemical companies has said that the entire cost of their research effort is supported by royalty income. Technology packages including both manufacturing know-how and patent rights are regularly sold by chemical companies both domestically and abroad for many millions of dollars, and such exchanges are the basis for much of the worldwide expansion in chemical technology. This growth will surely continue, and patent rights will be one of the most important factors.

Research and development and technology presently abound in American industry, universities, and government agencies. The specific organizations can give information about their own technology available for licensing in patent and/or trade secret form. The *Official Gazette* of the U.S. Patent and Trademark Office, Washington, DC 20231, routinely publishes lists of patents that are available for licensing from various government agencies, not-for-profit research institutions, universities, or specific companies.

11

The Employed Inventor: Assignments and Employment Agreements

Few chemists and chemical engineers work alone. Most are employed by industrial firms that are able to provide the elaborate, expensive equipment needed to practice their profession, the large sums of money necessary for manufacturing facilities, and the sales and distribution organizations that make the commercial application of their research and development results possible. Those not in industry are employed by colleges or universities, research institutes, consulting firms, and various government research laboratories. The lone inventor in the chemical professions, although a special and valuable member of society, is rare.

Assignments

For more than a century, U.S. law has required that patents be granted to the individuals who actually have made inventions or to their legal heirs if the inventor should die or become incapacitated. The property right represented by a patent can be transferred just as can any form of property, and it is customary for employed inventors to transfer ownership of their patents to their employers by means of a contract known as an "assignment". The inventor is the "assignor", and the employer is the "assignee".

Contracts to transfer property rights are expected to represent a fair exchange of some sort. There must be "consideration" of some value in return for the property being transferred. The consideration in the case of the employed inventor is the fact of employment, salary, provision of facilities and materials needed to carry out research and development, and the opportunity to do meaningful work. Sometimes an assignment contract will also include payment of a nominal sum of money, typically $1, to

1997–4/91/0105/$06.00/1
© 1991 American Chemical Society

avert any question about whether there is consideration, but the need for this practice is doubtful, and it is not as common as it once was.

The actual assignment paper is a legal document that identifies the patent application, the parties making the agreement, and in some cases the consideration. It is signed by the inventor(s) and a witness or a notary public. Usually it is filed with the application or soon after the application is submitted to the U.S. Patent and Trademark Office (PTO), although this is not mandatory. If the assignment is recorded in the PTO while the application is being prosecuted, the patent as printed will show the assignee, and this is the common situation. Sometimes the assignment is filed too late to appear on the patent, and in that case only the inventor(s) will be named on the patent. This practice does not necessarily mean that the patent is not assigned, and if ownership of the patent needs to be determined, an inquiry to the PTO will be necessary.

Ownership of a patent can be further transferred by any person or legal entity, as a corporation, that has acquired it from the inventor. In this case the transfer contract is known as a "mesne assignment", the adjective mesne being derived from the middle French *meien*, meaning intermediate in time of performance. This expression is occasionally seen in the heading of patents. Such assignment of rights to a patent beyond that made when the patent originally issued is not legally effective unless it is recorded in the PTO.

Ownership of Patents; Shop Rights

U.S. federal and individual state statute and case law recognizes that an employer is the rightful owner of the work output of those he employs to study and to improve the technology of the business concerned. The employee is "hired to invent". Inventorship is always decided according to federal law. However, ownership may be determined by federal or state law. Generally, professional "exempt" employees work full time, 24 hours per day, for their employers, in contrast to "nonexempt" employees such as laboratory technicians, plant operators, clerical personnel, and so on, who are employed for limited periods, usually 8 hours per day. The terms "exempt" and "nonexempt" refer to the status of employees under the Federal Labor Relations Act, which regulates working hours and pay practices for nonprofessional personnel, referred to as nonexempt. Professionals are usually exempt from these regulations.

Thus the professional employee is generally considered to owe the employer rights to all inventions that he or she may make during the time of employment. A question arises if the employee makes an invention that is outside the particular field of activity for which he or she is hired. Most employers are willing to release the employee from any obligation to assign the rights to such inventions, but the employer may choose to retain a "shop right" to use the invention if it appears that it may have some application in the business. In the chemical industry, an improved valve that the employer would not normally manufacture or sell but that could be useful in process equipment might be such an invention.

The concept of shop rights derives from the recognition that employees who are not hired to make inventions, or those who make inventions on their own time or outside their stated assignments, are often doing so by using the facilities provided by the employer or on the basis of the experience gained during their employment. The employee has the right to patent and to exploit the invention, enforcing the patent rights against any third parties who may want to use it, but the employer has a shop right to use the invention without paying any tribute to the employee. If the invention is such that it would have no application in the business—for example, an improved lawn sprinkler—many employers in chemical industries will waive all rights, leaving the employee free to exploit the invention himself.

Employment Agreements (Confidentiality and Trade Secret Agreements)

Most employers require their technical employees to sign an agreement that spells out their obligation to assign patent rights to any inventions they make during their employment. The agreement usually also includes a statement of the obligation of the employee to safeguard trade secrets and other confidential information, and in addition it may restrict any use of confidential information after employment is terminated. Such agreements are not usually required of people working in nontechnical aspects of chemical industry. Laboratory technicians are in some cases asked to sign invention agreements, and this practice is becoming more common because the training and responsibilities of such employees are continually being raised to more sophisticated levels.

Under the terms of the usual agreement, the employee accepts an obligation to disclose all inventions that he or she may make to the employer, to assign all patent rights, and to cooperate in obtaining a patent if the employer decides to file an application by providing the necessary information and signing any necessary papers, such as application forms, affidavits, and assignments.

The time to negotiate specific terms in an employment agreement is prior to signing the agreement and prior to starting employment. The following are negotiable terms for consideration: particular background areas, specific patents, or patent applications reserved as background rights for the new employee, etc. Some states (for example, California, Minnesota, and North Carolina) have enacted statutes that limit the scope of the patent rights that an employer can require to be transferred by the employee. Usually the limitations require that the assignment of patent rights is proper for inventions that are within the company area of business, are part of the employee's scope of responsibility, and have been conceived or reduced to practice by using company funds, time, information, and facilities.

Furthermore, an employment agreement should be signed by both the employee and the competent company representative no later than the first day of employment. Employment agreements signed later, during the course of normal employment, may be unenforceable for lack of sufficient (or adequate) consideration on the part of the new employee. Competent legal counsel can explain the effect of the law in specific jurisdictions.

A problem sometimes arises between employee and employer when the employer chooses not to seek patent protection for an invention. Other than the reasons discussed in Chapter 4 on the decision to file a patent application, the employer may simply decide that the invention is not of enough interest to justify the cost of prosecuting an application. Resolution of this problem must be a matter of good will and trust on the part of all parties involved. Sometimes an employee in this situation will ask for a release to exploit the invention personally, and many employers are willing to grant such a release unless disclosure of otherwise valuable proprietary information might result from publication of a patent.

The termination provisions of employment agreements may present another problem. Many such agreements place no obligation on the employee after termination, but some have provisions requiring assignment of inventions for some period of time after the employee leaves the company, on the theory that the idea for an invention may have been generated during the period of employment or is based on proprietary information. The law is unsettled on what time limits are enforceable. Those considering job offers should take such requirements into account in deciding where to accept employment, if for no other reason than the likelihood that such a restraint could limit opportunities for further employment in case a change of employers becomes necessary.

Compensation

Salary and other rewards such as bonuses are seldom addressed in employment agreements. In the United States, employed inventors usually do not receive large compensation for patentable inventions beyond normal salary, although some companies have schedules of special payments associated with patent activity. Fredrik Neumeyer, a Swedish consultant on industrial property law, has published an extensive case study of the practices of American companies in this area. One company has paid $200 for each invention disclosure chosen for filing of a patent application, $500 to each inventor who has had five U.S. patents issued in his or her name, as well as special awards of $1000 or more for the most valuable inventions. Another company has a point system that assigns value to patents and rewards inventors in amounts that have been as great as $50,000. Some companies have special compensation plans that reward exceptional contributions depending on the profits or savings that result, patents being only one factor in the judgment. There is no unanimity in the practices of the American chemical industry on special compensation related to patents.

On the other hand, many employees are hired by a corporation for the specific purpose of inventing. Some conflict may occur within the company if a specific class of employees is singled out for special treatment for their contributions. In some U.S. companies an invention incentive (or compensation) award of any amount may cause a lack of communication and disrupt the work place environment.

On balance, a majority of U.S. employed inventors apparently believe that they are properly compensated by their employer for inventions and patents within the scope of their employment. The American Chemical Society's Patent Committee survey reported by Howard J. Sanders in *Chemical and Engineering News*, May 26, 1980, pages 32–40, is instructive on this point.

In many European countries, as well as in Japan, employers are required by law to share with the inventor any profits resulting from patented inventions. West German law on this subject is best known. A formula that is provided assigns a weight to the level of responsibility and to the work assignment of each patentee, those at the highest organizational levels being given the lowest weight factors. Appropriate rewards depend on the estimated value of the invention over a limited time, the amounts paid being some fraction of the royalties that might be required for the right to practice patents held by an independent inventor or another company. In most cases, agreement on payment levels is reached by negotiation between the patentee and the employer, but an arbitration body in the German Patent Office considers the case if agreement cannot be reached, and appeal to the Supreme Court is possible. Relatively few cases go to arbitration, and very few go to the German Supreme Court.

In the Soviet Union the government pays rewards up to 25,000 rubles to inventors whose contributions are used in state enterprises. This system is discussed further in Chapter 15.

The great variety of systems for compensating inventors that are being tried suggests that the most effective method may not yet have been developed. However, they all share the common goal of attempting to stimulate creative invention as evidenced by patents, thus implementing the famous quotation by Abraham Lincoln, "The patent system added the fuel of interest to the fire of genius." The principle involved is expressed in the American Chemical Society's Professional Employment Guidelines, which say "Extraordinary contributions to patentable inventions, trade secrets, or know-how should be compensated by specific rewards commensurate with the value of the contributions to the employer" (Section II, paragraph 9).

Trade Secrets and Confidential Information

As already stated, nearly all employment agreements require the employee to protect and to keep confidential trade secrets and other information of a confidential nature. Trade secrets are protected by law in any case, as discussed in Chapter 12, but employers want to have the obligation clearly before the employee. Some agreements place a time limit on the obligation, on the theory that trade secrets have a limited useful life, and some attempt to limit the right of one who has left employment to compete in the same field or to take employment in a similar capacity to what he or she held previously. Enforcement of such agreements is possible in clear-cut cases where unfair use is being made of secret information, but the general trend of the law is that people may not be deprived of the right to make a living by using the general experience gained in their employment. Time limits on agreements not to compete of no more than a year or two are sometimes accepted by the courts, but a solid case must be made by the employer that trade secrets are truly involved in the situation.

Confidential information is in essence any internal written information such as notebooks, research reports, meeting minutes, memoranda, compilations of data, process details, and market research studies. The test is whether the information has been made publicly available in any way by the employer. If not, it is confidential, and the employee owes the employer care in its handling. This is a matter of ethical standards as well as written employment agreements and needs little discussion. The employee is of course obliged to return all such information when leaving employment.

Formal agreements are less common for those employed in academic or government positions, but the law and the ethical considerations that apply are comparable. Those who make inventions while employed by such organizations should make sure that they understand the rules and practices that apply to their particular situations to avoid potential conflicts with their obligations.

12

Copyrights, Trademarks, and Trade Secrets; Design and Plant Patents

Patent rights are one form of what is generally called "intellectual property". Copyrights, trademarks, and trade secrets are considered to be other forms of intellectual property, and each of these categories has its own body of statutory and case law. The legal profession includes those who specialize in each of these fields. The complexities of each of these types of property are beyond the scope of this book, but the chemist and engineer need to understand certain basic principles.

Copyrights

The U.S. Copyright Office of the Library of Congress is very helpful in providing the forms necessary to file a copyright application. The provisions of Article 1, Section 8 of the U.S. Constitution quoted in Chapter 1 provide the basis for our copyright system, just as they do for the patent system. The social purpose is much the same in each case—to encourage creativity and the advancement of knowledge by giving those who make intellectual contributions an "exclusive right to their writings and discoveries" for limited times. In copyright law this can include literary, musical, or artistic works. Protection is provided for a considerably longer time for copyrights than for patents. The Copyright Law of 1976, adopted after several years of consideration by the Congress, provides copyright ownership for the lifetime of the author plus 75 years. The earlier U.S. law provided a term of 28 years, renewable for a further 28 years. The present provision has become the generally accepted term for copyrights in most countries that are parties to the Berne Convention on Copyrights. It recognizes that the value of written works often extends over a longer time

1997–4/91/0113/$06.00/1
© 1991 American Chemical Society

than that of technological advances and that the placing of written works in the public domain is not as generally valuable to society as it is for patented inventions.

It is not essential to register written works for formal copyright protection in order to have at least some degree of protection against the copying of one's works by others. Unpublished writings are subject to common law rights that can be enforced in courts of equity. The copyright of a published work is created by committing the authored expression to a tangible form. The Copyright Office is a branch of the Library of Congress, and the procedure for obtaining a copyright registration is quite simple: An application is submitted with a modest fee and two copies of the work. A generally cursory examination is made, and the copyright is immediately in force.

The Copyright Office will provide on request a variety of bulletins describing their procedures and giving information about the classes of subject matter that can be registered. These classes, as well as those not subject to copyright, are shown in List 12-1.

Copyrights differ from patents not only in the ease by which they are obtained but also in the kind of protection provided. Where a patent gives the right to exclude all others from using the same invention, a copyright protects the holder only against those who knowingly copy or use the same form of expression as that of the author. Another author who independently creates a work of the same or nearly the same content and form has an equal right to obtain copyright protection. The copyright, in other words, protects against copying, especially with respect to the unique, original form of expression of the author. The novelty of the subject matter is not as important as it is with patents. Copyright infringement actions that try to prevent use by others of the sense of a copyrighted work usually succeed only if substantial portions of a copyrighted work in the author's exact language or form have been copied into another work.

The uses that are controlled by copyright include reproduction and distribution and public performance of dramatic or musical works for profit. Private use, loan, or sale of a personal copy to another person and of information gained from the work are not restricted by copyright. Uses such as photocopying of documents and distribution of program material by cable television networks have been the subject of extended discussion. The Copyright Law of 1976 allows "fair use" of copyrighted material, that is, copying of limited portions of a work for purposes such as news reporting, criticism, teaching, scholarship, or research. Single photocopies may be made of material such as a journal article, but systematic reproduction of multiple copies, as by a library, is forbidden. Licensing arrange-

---

List 12-1. Classification of Copyright Objects

Those subject to protection:

1. Books
2. Periodicals and newspapers
3. Oral talks–lectures and sermons
4. Dramatic compositions
5. Musical compositions
6. Maps
7. Works of art
8. Reproductions of works of art
9. Drawings of scientific character
10. Photographs
11. Prints and labels
12. Photoplays
13. Motion pictures other than photoplays
14. Sound recordings
15. Masks for semiconductor manufacture

Those not subject to protection:

1. Ideas
2. Systems, methods, plans
3. Titles by themselves
4. Laws and opinions of courts
5. Government publications

---

ments between publishers and libraries will no doubt be negotiated, in the way radio and television stations are licensed to use musical works.

To hold a copyright under the law, copyright notice must be included in the work at the time of the first public distribution, in the way prescribed as follows: The mark © and the words "Copyright by", the author's or publisher's name, and the year must be printed at a particular location. This notice should appear on every copy. Such publication secures the copyright. Deposit of two copies of the published work in the Copyright Office together with an application and the required fee provides the registration needed before an infringement suit can be instituted.

An international convention provides for reciprocal recognition of copyrights between signatory countries. Most of the major countries of the world, including the Soviet Union, are members.

Trademarks

Trademarks are like a badge or brand identifying the origin of a product or class of products, service marks, collective marks (as for a membership organization), or quality certification marks. Product trademarks should be distinguished from trade names, a more general term used to identify a producer or a commonly available product. Trademarks can be valuable property, especially when they identify products or services that have achieved a reputation for quality and reliability. Trademark owners put considerable effort into maintaining the uniqueness and identity of their marks with the particular products involved because careless use of trademarks can result in the loss of their proprietary value. Some of the best-known examples of such loss are the names cellophane, aspirin, escalator, and thermos, all of which were declared by courts to have become so commonly used as generic terms for those products that they could no longer be maintained as the unique property of the original owners.

Trademarks are protected by federal law under the Lanham Act of 1947. Unlike copyright, which is governed only by federal law, some states also have their own trademark systems. An essential requirement of U.S. law is that the trademark distinguish the source of the associated good or service from other sources. It is preferred that they be coined or fanciful words that are not likely to cause confusion with regular English-language words or as to source, origin, affiliation, or sponsorship in view of existing trademarks of other companies. The trademark system is administered by the U.S. Patent and Trademark Office (PTO), which will supply information on rules and procedures just as for patents. A force of trademark examiners studies applications for the registration of new marks to determine whether these marks meet the rules and whether they have ever been registered before. A classification system analogous to that used for patents is used to aid in this search. Marks similar to those used before can sometimes be registered if they apply to a class of products different from that of the earlier registration and if they are not likely to be sold in the same trade channels. A trademark registration can be obtained on the basis of either good faith use in interstate commerce or a bona fide intent to use the mark in the future in interstate commerce. The fact of commercial use provides a degree of protection, although many trademarks are simply used without registration. Registration provides rights in addition to those provided by mere use; however, marks considered important are usually registered.

A special aspect of trademark registration is that application to register a mark can be opposed by others who believe that the proposed new mark might cause confusion as to the source of the goods. Applications to register trademarks are published in a special edition of the *Official Gazette*, and commercial firms have staff people who regularly study this publication to watch for marks that might conflict with their own. The PTO has an appeal board that considers appeals from decisions of the examiners and oppositions. Decisions of the Trademark Appeal Board can be further appealed to the Court of Appeals of the Federal Circuit, just as can patent decisions.

Employed technical people must use trademarks properly to protect this valuable property. Proper use of competitors' trademarks is also advisable, to encourage reciprocal consideration by all. These are the important points in proper use of trademarks:

1. The trademark should be *distinguished* from accompanying text in some manner. This can be done by the use of quotation marks around the capitalized word, by use of all capital letters, by italics, or by other special forms.

2. The trademark should be *described* by linking it with the proper generic name of the product at least once, as KODAK cameras. A trademark is an adjective and should not be used as a noun. Trademarks are often used as nouns in ordinary conversation, but this can weaken the value of the mark and possible lead to its loss as a property. It is especially damaging to use a trademark with the incorrect generic term. For instance, "fabric of DACRON polyester fiber" is correct, but "DACRON fabric" is not because DACRON is the trademark for a particular polyester fiber; it is not a trademark for a fabric. This sort of trademark misuse is widespread, but technical people should be careful to avoid it because of their special responsibilities for protecting industrial property.

3. A registered trademark should be *designated* as such in commercial practice, as in labels, letters, and sales literature, by the mark ®, or by the notation Reg. U.S. Pat. and Trdmk. Off.

Similarly, service marks (™ or ®, for example, MacDonald's) identify the source of the service.

Trade Secrets

Patent law, trademark law, and copyright law are all ultimately governed by and decided by federal law, which is essentially uniform throughout the country. On the other hand, trade secret law is governed by each individual state's law. This situation can pose a problem for corporations having research or development facilities in a number of states. Because each state must look to its own statutes and historical case law to interpret the law, identical fact situations in different states may produce completely different (and even opposite) legal results.

A number of states have adopted a Uniform Trade Secret Act in an attempt to provide more uniform legal decisions in this area. States that have adopted this act include, for example, California, Connecticut, Delaware, Indiana, Kansas, Louisiana, Minnesota, North Dakota, Washington, and Wisconsin.

Trade secrets have been defined by the American Law Institute as follows: A trade secret may consist of any formula, pattern, device, or compilation of information that is used in a business and that gives the business owner an opportunity to obtain an advantage over competitors who do not know or use it. It may be a formula for a chemical compound; a process of manufacturing, treating, or preserving materials; a pattern for a machine or other device; or a list of customers. A trade secret is a process or device for continuous use in the operation of the business. Generally it relates to the production of goods, as, for example, a machine or formula for the production of an article. It may, however, relate to the sale of goods or to other operations in the business, such as a code for determining discounts, rebates, or other concessions in a price list or catalogue; a list of specialized customers; or a method of bookkeeping or other office management.

In some cases, when an invention has been made, a patent application is not filed for reasons such as the difficulty of enforcing a patent if obtained or a judgment that the disclosure necessary to obtain a patent would reveal even more valuable but unpatentable aspects of the commercial process. In such cases it may be decided to operate the invention as a trade secret.

Relying on secret use of an important improvement in a process or product has advantages and disadvantages. One advantage is that a secret has no time limit. Trade secrets are protected by state laws, and so long as the secret is maintained, legal action can be taken to protect such rights. Another advantage is that operating in secret, if feasible, avoids revealing to competitors information that might stimulate a successful effort to develop alternative ways of getting similar results.

The disadvantages relate mostly to the difficulty of keeping a secret. Technical personnel are highly mobile in our present society. It is difficult to prove in a court action the secret character of information and the adequacy of measures taken to maintain the confidentiality of proprietary information. Unscrupulous competitors do resort to unfair means to discover trade secrets. The holder of a trade secret has no legal rights against a competitor who develops the same information independently or by "reverse engineering". Finally, a competitor may discover and patent the same invention. A patent owner has exclusive rights against all others, but the trade secret owner can take action only against persons who violate a trust by revealing the secret or by using it in competition with a former associate or employer. Legal actions against those who start businesses in competition with a former employer on the basis of confidential knowledge sometimes recover damages or result in injunctions against the operation of the new enterprise, but the secret can be forever lost as a result.

Great care and effort must be made if a trade secret is to be successfully maintained. Document and material controls must be maintained, access to secret plant operations must be carefully restricted, precautions against industrial espionage must be taken, and employee agreements and termination procedures must be handled with care and thorough understanding of the laws involved, which differ from state to state.

The sharing of information for the purpose of evaluating its usefulness and joint development programs between industrial concerns are frequently undertaken with the protection of written agreements that call for keeping the shared information confidential for a limited time. Such agreements often include provisions with respect to ownership of any patents that may come from the joint effort. Patents are the best protection for important findings, but operation as a trade secret is a valid course under the right circumstances.

One of the longest held U.S. trade secrets is the formulation and process to produce Coca-Cola®.

Design Patents

The U.S. patents that have been discussed throughout this book are sometimes called utility patents. U.S. law also provides for two other types: design patents and plant patents. Design patents protect the design of articles of manufacture if the design is original. The emphasis is on the ornamental aspects of the article, as shown by a drawing. The grace period during which an application for a design patent can be filed after public disclosure is 6 months versus 1 year for utility patents, and the term is 14 years. Until recently the term depended on the fee paid by the applicant. Fees are modest, and the prosecution is less complex than for utility patents. Nevertheless, a patent for a distinctive design that meets extensive public acceptance can be a valuable property. Articles such as furniture, containers, toys, instruments, and structures are often the subjects of design patents.

Many countries have somewhat analogous types of patents, variously known as "petty patents", "utility models", or in Germany "Gebrauchsmuster". Examination and fees are minimal, and relatively short-term protection is provided, in keeping with the limited commercial life of most articles that depend on unique design for their appeal.

Plant Patents

Plant breeders as well as amateur agronomists frequently discover or deliberately develop new and distinct mutant or hybrid varieties of plants that have improved productivity, resistance to disease, or ornamental value. These new varieties may be patented. Tuberous plants, for example, potatoes and Jerusalem artichokes, are specifically excluded from plant patent protection. The patent grant conveys the right to exclude others from asexually reproducing the plant—that is, by propagation of cuttings or by grafting. Plant patents are usually illustrated with a color photograph. Copies cost $10.00.

Sexually reproduced plants can also be the subject of 17-year patentlike protection by means of "Certificates of Protection" issued by the U.S. Department of Agriculture under the Plant Variety Protection Act. After filing in a foreign country, applicants are allowed 5 years for filing in the United States because of the lengthy "grow-out" tests necessary to determine the characteristics of the new plants involved. A fee of $500 is charged, and a sample of viable seed of the plant variety for which protection is sought must be submitted with the application.

13

Recent Biotechnology-Related Patent Law

An exhaustive discussion of biotechnology patent-related matters is beyond the scope of this book. A brief summary is provided regarding general aspects of biotechnology, monoclonal antibodies, hybridomas, and diagnostics. A list of supplemental readings is provided in the Bibliography.

End Products Produced by Fermentation

The U.S. Patent and Trademark Office (PTO) has long granted patents having claims directed to methods for producing useful end products (for example, antibiotics) via a fermentation process. The fact that a microorganism was involved in the process did not keep the process from being patentable, provided that the microorganism was available to the public. If the microorganism is not available to the public, it can be made available by placing a culture of the microorganism in a public depository such as the American Type Culture Collection (ATCC) in Rockville, Maryland. A number of depositories are recognized pursuant to the Budapest Treaty (1973) on the International Recognition of the Deposit of Microorganisms for the purpose of Patent Procedure; these depositories will maintain deposits of various biological samples for patent purposes. A deposit made in a recognized (recognized for the Budapest Treaty purposes) depository in one country will be considered to be a suitable deposit for patent purposes by the other countries adhering to the Budapest Treaty.

NOTE: This chapter was written with Kenneth L. Loertescher, Esq., The Dow Chemical Company Midland, MI 48674.

1997–4/91/0121/$06.00/1
© 1991 American Chemical Society

The end products of fermentation processes also have long been considered to be patentable subject matter. Any material isolated from a fermentation broth (or isolated from any other biological sample) is patentable per se; provided, of course that it meets the basic patentability requirements of Title 35 of the U.S. Code, Sections 101, 102, 103, 112, etc. (35 U.S.C. 101, 102, 103, 112, etc.) (that is, is novel, useful, and unobvious, etc.). The claims to these materials are similar to any other claims directed to a chemical compound; that is, the compound can be depicted by a chemical formula and/or named and claimed in the same manner as any other chemical compound. In instances where the biologically produced material itself can still be claimed; however, it will have to be characterized sufficiently, for example, by specifying its melting point, chemical composition, solubility, biological activity, etc., so that it is clear that the material being claimed is different from and is patentable over compounds known in the art and in nature. Secondary metabolic products are the epitome of patentable subject matter because the secondary metabolic product(s) that are produced depend on a variety of factors, including the microorganism used and the conditions and substrate employed; thus, often a strong argument can be made that a previously unknown and useful secondary metabolite is unobvious over other compounds known in the art.

Many commercially viable materials that have been obtained via fermentation processes have been patented; the avermectins and related compounds are examples of compounds that are produced via a fermentation process, have been patented, and have significant commercial impact.

Natural Products

A more nebulous class of materials is referred to as "natural products" for lack of better terminology; these compounds are isolated from a tissue source. The PTO will grant claims directed to natural products (if the basic patentability requirements are met). However, it is often difficult to demonstrate that these materials are unobvious. Generally, the natural products are present in only very small quantities in the tissue source from which they are obtained (thus explaining why they have remained unknown) and usually require a significant degree of human effort to isolate in relatively pure form. The PTO usually requires that the claims be worded so that they are directed to the product in a "substantially pure" (or "essentially pure" or similar language) form in order to distinguish the claimed subject matter over the material as it exists in nature. The PTO has frequently tried to reject claims directed at biological inventions on the basis that they were products of nature and thus not patentable subject matter; however, the courts have routinely, at least in the recent past when otherwise patentable invention was involved, overturned this rejection when it has been decided on appeal.

The landmark decision involving the interpretation of the U.S. patent laws was reached in 1980 in *Diamond v. Chakrabarty,* Vol. 206 *U.S. Patent Quarterly* page 193. The microorganism that Chakrabarty sought to patent was not a naturally occurring microorganism; it was a bacterium from the genus *Pseudomonas* that had been genetically manipulated so that it contained "at least two stable energy-generating plasmids" (not previously found together in a single microorganism), and each of the plasmids provided a separate means to degrade different multiple components of crude oil. The U.S. Supreme Court held that a microorganism could be patentable subject matter (that is, that Chakrabarty's microorganism was a "manufacture" or "composition of matter" falling within the definition of patentable subject matter under 35 U.S.C. 101). The court indicated that Chakrabarty's microorganism was not a naturally occurring composition of matter and was a product of human ingenuity having a distinct name, character, and use.

Since the Chakrabarty decision, the PTO has taken a fairly liberal approach to granting patents claiming "living things". After the Chakrabarty decision, the PTO complied with a 1979 decision of the Court of Customs and Patent Appeals, *In re Bergy, Coats, and Malik,* Vol. 201 *U.S. Patent Quarterly* page 352, and has allowed claims directed to microorganisms that although naturally occurring, need some degree of human intervention in order to place them in a usable form; the microorganism claimed by Bergy et al. was isolated and found to produce the antibiotic lincomycin under the proper fermentation conditions. Thus, if a previously unknown and unobvious microorganism is isolated from the sample and if the isolation involves a sufficient degree of human intervention in order to obtain a pure culture of the microorganism, a claim directed to the novel microorganism can be made. When the microorganism is naturally occurring, that is, simply isolated from a sample, the PTO generally requires that the claim be drafted so that the claimed subject matter is differentiated over naturally occurring materials. Thus, a frequently used phrase in claim language is "a biologically pure culture of X microorganisms".

Of course any microorganism that is patented must also be useful, for example, used in a fermentation process to produce an antibiotic (or other desirable material) or useful in a process to selectively oxidize a particular substrate to give desired products. Other aspects of the invention can also be claimed, provided that the particular subject matter that is claimed meets the requirements for patentability. In the case of the production of a novel and unobvious antibiotic by a novel and unobvious microorganism, the process for obtaining the microorganism, the process used to make the antibiotic itself, the compositions containing the antibiotic, etc. can be claimed. Even when the microorganism itself is not patentable (that is, the microorganism is known), there still may be patentable subject matter; for example, a known microorganism might be used to produce new, useful, and unobvious compounds (the compounds and the process of using the compounds could be patented) or the known microorganism might be used in a new, useful, and unobvious process (the process could be patented).

When the microorganism (whether naturally occurring and isolated, or not naturally occurring and obtained by standard genetic manipulation or otherwise) is unavailable to the public, a culture of the microorganism will have to be deposited in a specified depository in order to meet the requirement set forth in 35 U.S.C. 112 (the patent application must teach one skilled in the art how to make and use the claimed invention). If the microorganism is unavailable, one skilled in the art cannot practice the invention. The PTO's position, and appropriately so, is that the procedures that ensure that the microorganism will be recovered cannot always be repeated. In these cases, a deposit of a culture of the microorganism at a depository will almost always be required. In a 1985 decision, *In re Lundak*, Vol. 227 *U.S. Patent Quarterly* page 90, the Court of Appeals for the Federal Circuit held that, under certain circumstances, a sample of a biological material referred to in a U.S. patent application (in *Lundak*, a cell line was involved) could be deposited *after* the patent application was filed. In any event, for U.S. purposes, the deposit must be made prior to the time at which the U.S. patent issues in order to make the invention available to the public.

However, it is generally wise to make the deposit of a microorganism (or other biological entity) *prior* to the time at which the patent application is filed, and the patent application as filed should refer to the deposit by including a taxonomic description of the microorganism deposited, its identifying name, the name and address of the depository, and the accession number which the depository assigned to the deposit. This prior deposit is required because the depositing requirements necessary to obtain patent protection in countries other than the United States are much more stringent. At this time, foreign countries require that the microorganism (or other deposited items) be deposited prior to or on the date at which the patent application concerning the microorganism is filed (in the United States and used later as the priority document within the Paris Convention).

Monoclonal Antibodies, Hybridomas, Diagnostics

A useful step in the purification of a hormone is to use a monoclonal antibody to specifically precipitate the hormone or to bind it onto a chromatographic column. An antibody is a protein that will specifically bind to a specific molecule, called an antigen.

Presented in Figure 13-1, left side, are the steps used in the production of antiserum that contains antibodies to a particular antigen, that is, a protein growth hormone. The antigen can be a protein, polysaccharide, virus particle, etc. A natural antigen usually contains a large number of potential antibody-binding sites. These sites are also termed antigenic determinants. The antigen is injected into a mammal such as a rabbit, sheep, goat, or calf. A number of antibodies are produced in the mammal's spleen to bind to one or more antigenic determinants. These antibodies can then be used as a reagent of mixed antibodies that bind to the antigen.

Conventional polyclonal antiserum presents a number of disadvantages. First, the antibodies are a mixture of a large number of different antibodies. These antibodies will bind with a large number of antigenic sites on the antigen. Cross-reacting antibodies are present, that is, those that also bind to other antigens, and therefore interfere with the use for which the antiserum is intended. Second, a particular antiserum cannot be duplicated, because each animal will respond differently to the same antigen. Thus highly specific high-titer antibodies are available only in limited amounts.

On the other hand, monoclonal antibody technology eliminates a number of the disadvantages of polyclonal antiserum. Figure 13-1, right side, shows the production of monoclonal antibodies. In 1974, Kohler and Milstein, in their Nobel Prize winning research, used cell fusion techniques to surmount the problem of lymphocyte nongrowth in culture. Their technique produced hybridoma cell lines. Spleen lymphocytes from a mammal immunized with the antigen are fused with mouse myeloma cell lines. Myeloma cells are immortal cells that will continue to grow so long as they are continuously fed. These hybrid cells have assumed an immortal

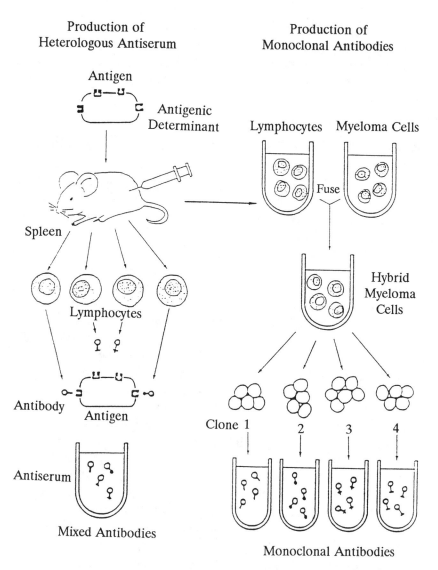

Figure 13-1. *Production of heterologous antiserum and monoclonal antibodies.
(Reproduced with permission from* Scientific American. *Copyright 1980
William H. Freeman.)*

life time and also produce the antibody specific to the lymphocyte. The
hybridoma cells are cloned and grow so that the groups of cells obtained
each produce a single monoclonal antibody. The disadvantages of this
monoclonal procedure are that in the fusion only about 1 in 200,000
lymphocytes forms a living hybrid, and of those hybrids only about 15 will
produce an antibody specific to the antigen. The cell lines must be carefully

screened to identify the clones that are producing the desired monoclonal antibody. The advantages of monoclonal antibodies are that they react with a single antigenic determinant, are available in an unlimited supply, can produce antibodies to a single molecule in a complex mixture, and can detect a component (antigen) in a mixture that is not detectable by conventional antisera in very small quantities.

In diagnostics, the monoclonal antibodies are particularly useful because of their selectively. Usually, the monoclonal antibody is covalently linked to a marker that is detectable by conventional analytical instrumentation. The now conventional enzyme-linked immunosorbent assay (ELISA) and other analytical techniques are based on this technology. Other examples include K. Rubenstein, et al., U.S. Patent 3,817,837, and *Scripps Clinic and Research Foundation v. Genentic Inc.*, Vol. 3 *U.S. Patent Quarterly*, 2d page 1481 (1987), U.S. Patent 4,361,509.

PTO Requirements

The PTO places considerable emphasis, at least with respect to naturally occurring biological materials and microorganisms, on the requirement for human ingenuity or human intervention. The degree or extent of human intervention required to obtain the biological material or microorganism can in large part dictate whether the PTO finds the invention unobvious.

The PTO has issued many patents directed to microorganisms. For example, various bacteria, yeast, fungi, and viruses have been patented, as well as particular genetic components such as plasmids and genes. In addition, patents have issued having claims directed to cell lines (for example, hybridomas), monoclonal antibodies, and other biological materials such as vaccines and proteins. In a 1985 decision, *Ex parte Hibberd, et al.*, Vol. 227 *U.S. Patent Quarterly* page 443, the PTO Board of Appeals and Interferences held that plants are patentable subject matter under 35 U.S.C. 101. The PTO is following this decision, and is now issuing patents claiming plants, seeds, etc.

On April 12, 1988, the PTO took the position that transgenic nonhuman mammals were proper patentable subject matter. U.S. Patent 4,736,866 for a genetically engineered mouse (P. Leder et al., assigned to Harvard College) is now being litigated in California.

The review of a biotechnology-related patent is the same as that for other patents. The scope of protection provided by a patent is defined by the claims, not what is taught in other parts of the patent. The scope of the teachings will have some impact on the interpretation of the claims and can be a factor (as is the prosecution of that patent application before the PTO) in a determination by a court (for example, in an infringement action) whether or not certain inventions that are not literally claimed are nonetheless deemed to be equivalent inventions covered by the claims. Of course, the teachings of the patent will be available to the public upon issuance of the patent and can have significant impact on the patentability of similar inventions made after the publication of the patent. Furthermore, having a patent in the United States allows the patent holder to exclude others from making, using, or selling the claimed invention in the United States only for the duration of the patent. It does not provide the patent holder with an absolute right to practice the invention because other patents may dominate or cover all or a portion of the invention.

The field of biotechnology has become an important commercial endeavor in just a short period of time. Many researchers believe it is so important that it would have proceeded whether or not patent protection was available.

Supplemental Reading

Beier, F. K.; Crespi, R. S.; Straus, J. *Biotechnology and Patent Protection: An International Review (1985)*; Organization for Economic Cooperation and Development, 2 rue Andre-Pascal 75775 Paris CEDEX 16, France. U.S. address, 1750 Pennsylvania Avenue, N.W., Washington DC 20006-4852; (202) 724-1857.

14

Changes
in U.S. Patent Laws:
1980–1990

The U.S. laws regarding patents that have been enacted by the U.S. Congress are found in Title 35 of the United States Code (often referred to 35 U.S.C). The rules promulgated by the U.S. Patent and Trademark Office (PTO) are found in Title 37 of the Code of Federal Regulations (often referred to as 37 C.F.R.). The *Manual of Patent Examining Procedure* (M.P.E.P.) describes in detail the procedures that the U.S. patent examiners use to examine and allow U.S. patents.

The U.S. patent laws have undergone significant changes over the past 11 years since the first edition of this book. These U.S. patent laws are summarized briefly in this chapter. However, existing patent laws are constantly changing and being reinterpreted; a registered U.S. patent attorney should be consulted regarding specific fact and law questions.

The Patent Act of 1981

1. The Court of Appeals for the Federal Circuit was created. The older Court of Customs and Patent Appeals became the Court of Appeals for the Federal Circuit (usually referred to as the C.A.F.C.). This is a panel of 12 judges (when sitting en banc) who hear and decide cases having specific federal jurisdiction. These areas include patent, trademark, copyright, and antitrust. The efficient manner in which the C.A.F.C. has operated over the past 10 years has been a strong factor in the strengthening of the entire U.S. patent system. Questions of patent law within the PTO Board of Appeals are appealed to the C.A.F.C. A U.S. District Court decision from any district is also appealed to the C.A.F.C. C.A.F.C. decisions are appealed to the U.S. Supreme Court.

1997–4/91/0129/$06.00/1
© 1991 American Chemical Society

Currently, Pauline Newman, a member of the ACS and a former member of the ACS Council, the ACS Patent Committee, and ACS Board of Directors, is a judge of the C.A.F.C.

2. Maintenance fees were imposed on U.S. patents, payable at $3\frac{1}{2}$, $7\frac{1}{2}$, and $11\frac{1}{2}$ years. If these fees are not paid, the patent expires and is not enforceable. Under certain narrow circumstances, a maintenance fee can be paid retroactively, and the patent will be reinstated.

3. The fees for filing and prosecution were changed to provide for lower fees for individuals and small entities.

Patent Law Improvements Act of 1984 (Public Law 98–6222 (H.R. 6268))

Joint Inventorship. The law with regard to joint inventorship (35 U.S.C. 116) was amended to govern the situation that persons may be joint inventors even though

- they did not physically work together or work on the invention at the same time,
- each did not make the same type or amount of contribution, or
- each did not make a contribution to the subject matter of every claim of the patent.

This change is particularly useful in corporate research where, prior to this change, multiple patent applications often had to be filed to provide the correct inventorship for each invention.

Prior Art. Rejections in patent applications were sometimes made under 35 U.S.C. 102(f) (that is, the applicant did not invent the claimed subject matter) through 35 U.S.C. 103. The basis for the rejection was that the invention claimed in an application of one "inventive entity" (an earlier first group of inventors) is obvious over the work of a different inventive entity (a second slightly different group of later similar inventors) if the first inventive entity had knowledge of the earlier work. This situation is common in corporate research organizations.

Under the present law, when both inventive entities (that is, different groups of inventors) are employed by the same company (or have an obligation to assign their patent rights to the same business or corporate organization) at the time the inventions are made, the rejection is not appropriate and can be overcome by timely filing declarations of the inventors or the assignee.

Statutory Invention Registration

A Statutory Invention Registration (usually referred to as a SIR) differs from a patent in that it is published by the PTO, as noted in the *Official Gazette*, without being examined. By requesting a SIR registration, the inventor(s) irrevocably waives the right to receive a U.S. patent covering that invention. The main benefit of a SIR is that it is a publication to establish prior art against another inventor for essentially the same subject matter.

Correction of Inventorship

Under the amendments to 37 C.F.R. 1.48, the addition or deletion of inventor(s) is more now easily achieved by

1. the filing of a petition with statements of facts verified by the originally named inventors,
2. the taking of an oath or a declaration by each actual inventor,
3. the payment of a fee as found in 37 C.F.R. 1.17(h), and
4. written consent of any assignee.

Rights and Requirements of Developing Patentable Inventions with Federal Funding

Two U.S. laws have profoundly changed the patent policy involving federal funding of research for an individual (independent) inventor, a not-for-profit research institution, a college or university, or a small business: 35 U.S.C. 200–212 [Public Law 96–517 (1981)] and 37 C.F.R. 401.1–401.16 [Public Law 98–620 (1984)].

To obtain title to a patent, the contractor under whose control the invention occurred needs to perform a number of required actions:

1. The contractor must elect to retain title to the invention by disclosing the invention to the funding federal agency within 2 months of the inventor(s)' disclosure to the contractor.
2. The contractor must elect (in writing) to retain title within 2 years of the initial disclosure.
3. The contractor must file a patent application prior to any statutory bar.

Failure of the contractor to perfect title may result in forfeiture to the government. Furthermore, the inventor(s) may then request retention of patent rights, created on a federally funded project.

The U.S. government retains a nonexclusive, nontransferable irrevocable paid-up license to practice the invention or to have it practiced on behalf of the United States.

The contractor can grant an exclusive or nonexclusive license for the full term or a portion of the patent (a maximum of 17 years).

The statutes just cited or a registered U.S. patent attorney can provide additional details about this specific area of patent law.

The Drug Price Competition and Patent Term Restoration Act of 1984

The Drug Price Competition and Patent Term Restoration Act (21 U.S.C. 355) was enacted on September 24, 1984. This law governs specific procedures toward gaining timely Food and Drug Administration (FDA) approval for the sale of generic versions of certain drugs. The act also provides under a rather complex (and sometimes ambiguous) set of rules for the partial restoration of patent protection lost due to FDA regulatory delays, for drugs. At this time, similar patent term restoration is not available for FDA approvals needed for diagnostic chemicals subject to Environmental Protection Agency (EPA) regulation (insecticides, rodenticides, fungicides, etc.).

15

Trends
in U.S. and World
Patent Law

United States Law

Nothing is so permanent as change, and patent law and procedures are no exception. Although the Patent Act of 1952 and its predecessor statutes have served us well, many believe that it is time for some modernization of U.S. law, at least partly along the lines recommended in 1966 by the President's Commission on the Patent System. However, changes in the law will have no significant bearing on the general principles that have been discussed in this book, and at this writing there is considerable uncertainty about when a new U.S. patent law may be adopted, despite many years of consideration by Congress.

Any new law, if enacted, may alter prosecution procedures and place greater emphasis on such requirements as disclosure to the examiner of prior art known by the applicant, but the factors that the working chemist and chemical engineer need to understand in making constructive and profitable use of the patent system will not change substantially.

A major pressure for modification of U.S. patent law derives from the undisputed fact that a high percentage of patents that become involved in litigation are found to be invalid by the courts. The proportion found invalid is variously reported as 40–60%, depending on the basis of the analysis. Former U.S. Supreme Court Justice William O. Douglas once made the cynical comment that the only valid patents are those that the Supreme Court had not been given a chance to consider. This view and interpretation of the statistics on the number found invalid must be tempered, however, by the obvious fact that only patents whose validity is

1997–4/91/0133/$06.00/1
© 1991 American Chemical Society

subject to at least some doubt are going to be litigated. No reasonable person or corporation undertakes the risk and expense of defending commercial operations against a charge of patent infringement unless there is a significant chance of winning the case, because there is substantial doubt about whether infringement has actually taken place or about the validity of the patent in question.

Most patents serve their purpose unchallenged because they in fact represent an inventive contribution. Many patents that are granted may seem trivial to the casual observer, but in a free society it is surely appropriate that the business decision on whether to apply for a patent should be left to the applicant and that the allowance of patent claims should not be subject to value judgments by patent examiners.

The following are the major areas expected to be considered for change in the next revision of U.S. patent law:

1. The term of U.S. patents is expected to be based on the filing date rather than on the date of grant as at present. This would make U.S. practice similar to that of most of the other countries of the world, as shown in Table 7-1. The rationale for having the term start to run when the application is filed is that it discourages deliberate delay in prosecuting patent applications that could result in unfair extension of the period of monopoly available to the applicant. The term will probably be 20 years from filing, with a possibility of extension for reasons beyond the control of the applicant.

2. There is interest in some form of opposition procedure that would allow third parties to submit prior art known to them that did not come to the attention of the examiner during prosecution. One strong argument for this procedure is the fact that the statistics on patent litigation show that one of the most common reasons for holdings of invalidity by the courts is the introduction of pertinent prior art of which the examiner was not aware during prosecution of the application. In cases where no new art is found, a much higher proportion of litigated patents is held to be valid.

3. Strong arguments favor some form of interparty proceedings. Present U.S. patent prosecution is entirely ex parte, meaning that only the applicant deals with the examiner in seeking to reach agreement on allowable claims. The procedure is confidential, and no information is available to the public until final grant of a patent. Many persons who are concerned with the strength of the patent system believe that the participation of a third party to represent the public interest in interparty proceedings would lead to issuance of a higher proportion of clearly valid patents. The third party may be a special force of legal personnel representing

the public interest in the U.S. Patent and Trademark Office (PTO), or it may be those who enter oppositions after public disclosure of the application, or both. It has been proposed that the PTO should have extensive subpoena powers to obtain information relative to patentability.

4. Under the Patent Law of 1981, inventors, contractors, and patent attorneys are under a high duty to inform the PTO of art material to the examination and prosecution of a patent application. Presently the U.S. Congress is examining a new law that would impose a broader and even higher duty to inform requirement that the examiner be informed of all art known to the applicant that is in any way pertinent to the claims sought. Court-developed case law makes such a requirement inherent in present patent practice—it is said that the applicant has an "uncompromising duty" of complete candor in relations with the examiner because of the ex parte nature of patent prosecution. PTO rules now make an explicit statement to this effect.

5. A new law may emphasize more strongly that the best way of practicing the invention known to the inventor at the time of filing the application be not only included but specifically pointed out in the disclosure. This law would put an end to any present tendency of patent applicants to obscure the preferred procedure.

6. A provision for deferred examination of patent applications is possible. The argument for this procedure is that many patent applications are filed for speculative or defensive reasons, and after a few years' experience the applicants lose interest in obtaining a patent. The cost of examining such applications could be avoided, and the technology would be in the public domain sooner. After a limited time, for example, 5 years, an application would be abandoned if examination were not requested by the applicant or by a third party. The application would automatically be published under such a procedure, probably 18 months after the U.S. filing or foreign priority date.

The advantages urged for these changes in patent law are that U.S. patents would be more difficult to obtain but would carry a much stronger presumption of validity than do those now issued. Ideally, there would be less uncertainty about the significance of issued patents, and the high cost of patent litigation would be reduced. The disadvantages of these changes would be that the cost of obtaining patents would increase because of the greater search and legal fees necessary to meet the more demanding requirements for disclosure of known art and for defense against third-party opposition procedures. This increased cost could inhibit the filing of applications by individual inventors and small enterprises, who have been the source of many of the most important contributions to technological progress. Preferably, a new U.S. patent law will strike a constructive balance that will both strengthen the patent system and continue to encourage creativity.

Harmonization of the World Patent Laws

No discussion of U.S. patent law is complete without mentioning the coming harmonization of the world patent laws. Just as the European Economic Community (EEC) will become a reality in 1992, the major industrial nations are busy with the harmonization of the world patent laws. It is not yet clear what impact this will have for U.S. patent law. In the give and take of the negotiations, the United States will be forced to reevaluate its present grace period of 1 year after public disclosure, secrecy for all U.S. patent examination, ex parte patent examination and prosecution, publication of the patent application 18 months after U.S. filing, and the "first-to-invent" system, among others. It is not yet clear what points of concession will be made or accepted by the other industrial countries.

Foreign Law

The patent laws of other countries are also evolving. Among changes that are meaningful to foreign patent applicants are the following. France is gradually adopting an examination system after many years of dependence on a registration system. Study commissions in Great Britain and Canada have considered modifications of the laws of those countries, and a new law has been proposed for Great Britain. Canada has considered changes in its compulsory licensing of patent rights in fields other than foods and pharmaceuticals, which are now subject to that requirement, but no action on a new Canadian patent law seems imminent at this writing. Canada has recently eliminated its "grace period" and will shortly adopt a first-to-file statute. Germany and Japan now allow claims to chemical compounds, where previously only the processes by which they are made could be patented. Mexican patent law has become much more restrictive, to force greater dependence on internal development and exploitation of new technology through shorter patent terms, more limited rights, and stricter working requirements.

Perhaps the most significant trend has been the move toward deferred examination systems, both for the reasons mentioned under the discussion of possible changes in U.S. law and to reduce the burden of huge backlogs of unexamined patent applications in some countries. West Germany, the Netherlands, and Japan practice deferred examination, with a time limit of 7 years during which the applicant or a third party may request examination. Australia has a deferral period of 5 years, which can be accelerated by decision of the patent office or on request by the applicant or a third party. Brazil has a deferral period of 2 years. Experience in some countries has borne out the expectation that a significant proportion of applications will be allowed to lapse without examination under this sort of system.

International Agreements

The Patent Cooperation Treaty (PCT) came into existence on January 24, 1978, to provide a number of advantages to U.S. and foreign patentees interested in the prosecution and obtaining foreign patents in one or more foreign countries.

The PCT is often described as a "holding pattern" that provides the potential patentee with a moderately low-cost route to keep foreign patent rights alive while obtaining more time to evaluate the foreign markets and determine whether the costly patent filing, translation, and foreign associate fees can be justified on an economic basis.

The PCT provides for the filing of a single application, usually in English, French, or German, to a Receiving Office (for example, at the U.S. PTO or European Patent Office) and the payment of a single filing fee. The need to file a translation of the patent application is also deferred.

Chapter I. Under Chapter I of the PCT, a U.S. patentee must file a PCT application within one calendar year of the U.S. filing date to gain the benefit of the earlier priority date. Errors in the PCT application, obtaining a U.S. foreign filing license, obtaining formal drawings, transmission of the record copy to the International Office in Geneva, Switzerland, and transmission of the certified copies of the priority U.S. patent applications usually occurs within 4 months of the PCT filing. An international literature search is performed and transmitted to the U.S. inventor. The inventor has an opportunity to amend the claims of the PCT application in view of the literature search. Usually about 6 months after the PCT filing, the application is published as a publication in the PCT Swiss Office, and is then available (disclosed) to the public.

Under Chapter I of the PCT, the applicant must file in the designated national countries (or in the European Patent Office, etc.) within 20 months of the first (earliest) priority filing date. Some countries at 20 months also require a translation of the application into the official language for the purpose of prosecution in the foreign country, prior to filing in the national country.

Chapter II. In July 1987, the United States became a member of and subject to Chapter II provisions of the PCT. Under this portion of the PCT convention, the applicant has an opportunity to extend the actual filing date in most of the national countries (or the European Patent Office) to 30 months from the earliest priority date. The applicant must file a demand, pay the necessary fees (currently about $600), and request a second search of the literature before the end of the 19th month from the earliest priority patent application. Optionally, the claims can be amended. It is also possible to obtain coverage in the EPO convention (12 countries) and in the Organization African de la Propriete Intellectuelle (OAPI).

It is not possible to cover all aspects of the PCT convention here. In view of the changes that are occurring within the foreign patent laws, it is imperative that the U.S. applicant consult with a patent attorney to determine the best foreign filing strategy considering the most recent changes in the laws.

As is shown in Table 15-1, the international agreements bind most of the industrial nations, and many of the emerging Third World countries. The PCT can be judiciously used to defer the national filings of most of the foreign patent applications (with their corresponding high filing costs and translation costs).

European Patent Convention

The European Patent Convention (EPC) of the European Patent Office (EPO) come into existence in 1978. The EPO has headquarters in Munich, Germany. The EPO countries are presently Austria, Belgium, Federal Republic of Germany (West), France, Greece, Italy, Liechtenstein, Luxembourg, Netherlands, Spain, Sweden, Switzerland, and the United Kingdom. (*See* Table 15-1.) An EPO patent can be prosecuted in English, French, or German. It is published as a publication 18 months after the earliest priority document. Examination can be deferred up to 7 years and requires an examination fee. Annual fees are necessary to keep the application in an active status. When the EPO application is allowed and issued, the separate national applications are filed with the corresponding filing fees, annuities, local associates, and translations into the national languages. If an EPO patent is granted, usually the national patent will issue without further examination.

Table 15-1. Member Countries of the PCT as of 1991

PCT	EPO–EPC	OAPI Countries
Australia		
Austria	X	
Barbados		
Belgium	X	
Brazil		
Bulgaria		
Cameroon		X
Central African Empire		X
Chad		X
Congo		X
Denmark		
Germany (East & West)	X	
Finland		
France	X	
Gabon		X
Hungary		
Italy	X	
Japan		
Liechtenstein	X	
Luxembourg	X	
Madagascar		
Malawi		
Mali		X
Mauritania		X
Monaco		
Netherlands	X	
North Korea		
Norway		
South Korea		
Rumania		
Senegal		X
Soviet Union		
Sri Lanka		
Sudan		
Sweden	X	
Switzerland	X	
Togo		X
United Kingdom (England)	X	
United States of America		

Soviet Bloc Countries

Countries of the Soviet Bloc also have vigorous patent systems. The interest of westerners most often centers on that of the U.S.S.R. because of its great economic importance. The Soviet Union is a member of the Paris Convention, and it is becoming increasingly common for U.S. companies to file applications in that country as well as other East European countries.

The U.S.S.R. grants both patents and Inventors' Certificates. The examination procedure for each is much like that of other countries. Foreign applicants generally request patents, and a State Committee is charged with studying new proposals for industrial development to ensure that foreign-held patents will not be infringed by the proposed activities. If they will, licensing arrangements are negotiated. Soviet nationals nearly always apply for an Inventor's Certificate, which, if granted, is the property of the state. The inventor is rewarded according to an established scale if the invention is used in a state enterprise. Such awards can be as much as 25,000 rubles, or about $30,000. A State Committee also makes decisions on foreign filing of patent applications on inventions by citizens, and U.S. patents of Soviet origin are issuing more frequently each year.

16

Representative
U.S. Patent Fees
and Payment of Money

This chapter gives the usual U.S. patent fees and other money paid as of January 1991. Increases in fees are inevitable as time passes.

If the inventor or assignee is a small entity, university, small business, etc., then the fees are one-half the usual fee.

The numbers 1.16 to 1.21 refer to specific sections of Title 37 of the Code of Federal Regulations.

1.16 National Application Filing Fees

    a.   Basic fee for filing each application for an original patent, except design or plant cases:

        By other than a small entity: $630.00

    b.   In addition to the basic filing fee in an original application, for filing or later presentation of each independent claim in excess of three:

        By other than a small entity: $60.00

    c.   In addition to the basic filing fee in an original application, for filing or later presentation of each claim (whether independent or dependent) in excess of 20. (Section 1.75(c) indicates how multiple dependent claims are considered for fee calculation purposes):

        By other than a small entity: $20.00

    d.   In addition to the basic filing fee in an original application, if the application contains, or is amended to contain, a multiple dependent claim(s), per application:

        By other than a small entity: $200.00

If the additional fees required by paragraphs (b), (c), and (d) are not paid on

1997–4/91/0141/$06.00/1

filing or on later presentation of the claims for which the additional fees are due, they must be paid or the claims canceled by amendment, prior to the expiration of the time period set or response by the U.S. Patent and Trademark Office (PTO) in any notice of fee deficiency.

1.20 Post-Issuance Fees

a. For providing a certificate of correction of applicant's mistake (1.323): $60.00

b. Petition for correction of inventorship in patent (1.324): $140.00

c. For filing a request for re-examination (1.510(a)): $2000.00

d. For filing each statutory disclaimer (1.321):

    By other than a small entity: $100.00

For maintaining an original or reissue patent, except a design or plant patent, based on an application filed on or after December 12, 1980, and before August 27, 1982:

e. In force beyond 4 years; the fee is due by 3 years and 6 months after the original grant: $245.00

f. In force beyond 8 years; the fee is due by 7 years and 6 months after the original grant: $495.00

g. In force beyond 12 years; the fee is due by 11 years and 6 months after the original grant: $740.00

For maintaining an original or reissue patent, except a design or plant patent, based on an application filed on or after August 27, 1982:

h. In force beyond 4 years; the fee is due by 3 years and 6 months after the original grant:

    By other than a small entity: $490.00

i. In force beyond 8 years; the fee is due by 7 years and 6 months after the original grant:

    By other than a small entity: $990.00

j. In force beyond 11 years; the fee is due by 11 years and 6 months after the original grant:

    By other than a small entity: $1480.00

k. Surcharge for paying a maintenance fee during the 6-month grace period following the expiration of 3 years and 6 months, 7 years and 6 months, and 11 years and 6 months after the date of the original grant of a patent based on an application filed on or after December 12, 1980, and before August 27, 1982: $120.00

l. Surcharge for paying a maintenance fee during the 6-month grace period following the expiration of 3 years and 6 months, 7 years and 6 months, and 11 years and 6 months after the date of the original grant of a patent based on an application filed on or after August 27, 1982:

By other than a small entity: $120.00

m. Surcharge for accepting a maintenance fee after expiration of a patent for nontimely payment of a maintenance fee where the delay in payment is shown to the satisfaction of the Commissioner to have been unavoidable: $550.00

1.19 Document Supply Fees

a. Uncertified copies of the PTO documents:
1. Printed copy of a patent, including a design patent, statutory invention registration or defensive publication document, except color plant patent or color statutory invention registration: $1.50
2. Printed copy of a plant patent or statutory invention registration in color: $10.00
3. Copy of patent application as filed (each 30 pages): $10.00
4. Copy of patent file wrapper and contents, per 200 pages or a fraction thereof: $75.00
5. Copy of PTO records, except as otherwise provided in this section, per page: $0.50
6. Microfiche copy of microfiche, per microfiche: $0.50
7. Copy of patent assignment record: $1.50

b. Certified copies of PTO documents:
1. For certifying PTO records, per certificate: $3.00
2. For a search of assignment records, abstract of title and certification, per patent: $12.00

c. Subscription services:
1. Subscription orders for printed copies of patents as issued, annual service charge for entry or order and 10 subclasses: $7.00
2. For annual subscription to each additional subclass in addition to the 10 covered by the fee under paragraph (c)(1) of this section, per subclass: $0.70

d. Library service (Title 35 U.S. Code, Section 13): For providing to libraries copies of all patents issued annually, per annum: $50.00

    e.   List of patents in subclass:
        1.   For list of all U.S. patents and statutory invention registrations in a subclass, per 100 numbers or fraction thereof: $1.00
        2.   For list of U.S. patents and statutory invention registrations in a subclass limited by date or number, per 50 numbers or fraction thereof: $1.00
    f.   Microfiche copy of patent file record: $6.00
    g.   Uncertified statement as to status of the payment of maintenance fees due on a patent or expiration of a patent: $3.00
    h.   Uncertified copy of a non-United States patent document, per document: $10.00
    i.   To compare and certify copies made from PTO records but not prepared by the PTO, per copy of document: $5.00

1.21 Miscellaneous Fees and Charges

The PTO has established the following fees for registration of attorneys and agents:

    1.   For admission to examination for registration to practice, fee payable upon application: $270.00
    2.   On registration to practice: $90.00

Glossary
and
Abbreviations

Abandon: To relinquish, explicitly or implicitly, a potential patent right. An application becomes abandoned by failure to respond to an office action within the required time, or by formal ("express") declaration. A patent right can be abandoned by simple inaction.

Absolute Novelty: A term used to describe the patent requirement in most of the countries of the world that prior public disclosure or sale anywhere in the world before filing of a patent application within a country subject to the Paris Convention will be an absolute bar to obtaining a valid and enforceable patent in these countries.

Affidavit: A sworn, written statement describing facts and supplementary arguments in support of patentability of a patent application.

Aggregation: An assembly of parts that do not cooperate to make a patentable improvement. This term is seldom used today.

Allowance, Notice of: Written notice to applicant that one is entitled to a patent after prosecution has been closed. Issue fee must be timely paid or the application will be abandoned.

Amendment: A change in any part of the application submitted to meet rejections, objections, or requirements made by the patent examiner.

Annuity: An annual fee required by many countries to maintain a patent in force. *See* Maintenance Fee.

Anticipation: Prior art that negates novelty. A composition, method, or structure that is essentially the same as the invention and that existed before the invention was made.

Appeal: A response to a patent examiner's action (usually a Final Rejection) to the Board of Appeals within the U.S. Patent and Trademark Office.

1997–4/91/0145/$06.00/1
© 1991 American Chemical Society

It usually includes a Notice of Appeal (*which also see*) and an Appeal Brief (*which also see*).

Appeal Brief: The actual document (and fee) timely filed within the U.S. Patent and Trademark Office setting forth the applicant's arguments, points of law, discussions, and conclusions.

Appeal, Notice of: A document (and fee) timely filed in the U.S. Patent and Trademark Office to reserve the right to file an Appeal Brief.

Applicant: Person applying for a patent; must under U.S. law be the inventor or inventors except in special circumstances.

Application, Continuation: A second (or subsequent) application directed to the same invention claimed in a prior co-pending application of the same applicant.

Application, Continuation-in-Part: An application filed during the pendency of an earlier application of the same inventor, the later application containing at least a substantial part of the disclosure of the earlier application together with additional matter not so disclosed. It is often referred to as a "c-i-p".

Application, Co-pending: A related application that is pending before or after another has been filed and before such other application has issued or become abandoned.

Application, Divisional: A later application derived from an earlier co-pending application by the same applicant, claiming a nonelected invention from the earlier application.

Application for Patent: Papers comprising petition, specification, drawings when required, one or more claims, oath or declaration, and filing fee whereby an applicant seeks a patent.

Article of Manufacture: An article of any derivation that is proper subject matter for a U.S. patent subject to 35 U.S.C. 101.

Assert: To assert a patent is to attempt to enforce the legal right to exclude others from practicing the invention claimed.

Assignee: One to whom the patent right is legally transferred.

Assignment: Transfer of all or limited rights under a patent.

Assignor: One who assigns a patent right.

Auslegeschrift: An examined German patent application, laid open to the public for review and for opposition.

Bar, Statutory: Circumstances that prevent a valid patent from being granted, for example, publication anywhere before invention or in the United States, 1 year or more, before a patent application in the United States. Publication, public use, or sale before the first application is a statutory bar in many foreign countries.

Biotechnology: The emerging field encompassed by genetic engineering, gene splicing, hybridomas, monoclonal antibodies, etc.

Board of Appeals: A board of senior patent personnel that hears appeals from adverse decisions of examiners upon applications for patents.

Brevet: A patent (French).

C.C.P.A.: *See* Court of Customs and Patent Appeals.

Certificate of Mailing, Regular Mail: For certain responses to the U.S. Patent and Trademark Office (for example, an amendment or payment of the issue fee), a signed certificate of mailing will be effective under 37 C.F.R. 1.8. The referenced response will be considered as timely filed as of the certified date.

Certificate of Mailing, Express Mail: For certain actions before the U.S. Patent and Trademark Office (for example, filing of a new patent application, a continuation, a divisional, or a continuation-in-part patent application), only express mail filing by certificate according to 37 C.F.R. 1.10 will be accorded the certified date of mailing.

C.F.R.: Code of Federal Regulations. Title 37 includes rules and regulations for patents, trademarks, and copyrights.

c-i-p: *See* Application, Continuation-in-Part.

Claim, Dependent: A claim including, by reference to another claim, all its subject matter, and containing in addition some further limitation or restriction.

Claim, Generic: A claim to a generic invention, usually including within its scope the subject matter of subordinate (more narrow) claims.

Claim, Independent: A claim that is usually generic and does not have a reference (or dependency) to another claim.

Claim, Jepson: A stylized form of a patent claim having a preamble of the aspects known in the art, followed by a linking phrase "the improvement comprising", and then a description of the claimed patentable improvement.

Claim, Markush: A claim containing the phraseology, "A member of the class consisting of . . . ", linked with the conjunctive word "and" or its equivalent. To be allowable, the members of the "Markush group" must have at least one common property that is mainly responsible for their function in the claimed relationship. *See* Markush Group.

Claim, Multiple Dependent: A claim that references (and is limited by) more than one preceding independent and/or dependent claim.

Claim, Process: A claim to a method invention or to the method aspect of an invention.

Claim, Product: A claim to the physical form of an invention or to an invention whose form is physical, for example, a chemical compound as opposed to the method of making it.

Claim, Species: A claim within the scope of, but not coextensive with, another claim in the same patent or application.

Claim, Specific: A claim setting forth details of an invention and usually to only one form thereof; may or may not be accompanied by a generic claim.

Colorable Deviation: A small change made from what is claimed in a patent solely for the purpose of avoiding literal infringement of the claim.

Combination, Patentable: A series of process steps, mechanical elements, or a mixture of materials that produce a desirable effect or result that is not an obvious summation of the effects of the different steps, elements, or materials.

Commissioner of Patents and Trademarks: The highest appointed official (by the President and confirmed by the Senate) of the U.S. Patent and Trademark Office. All communications to the Office (or to the Patent Cooperation Treaty) are addressed to the Commissioner.

Composition of Matter: One of the statutory classes of invention in which the substance, not the form or shape, is the inventive subject matter.

Comprising: Including. A claim that recites "comprising" certain elements is not avoided ordinarily by the addition of another element.

Computer Programs: Recently held to be patentable subject matter when included in a process of commands, as one of the process steps. An action step after the step requiring the computer program step is required.

Conception: The mental formulation of an invention.

Conflict: A contest in the Canadian Patent Office between rival inventors claiming the same or substantially the same invention.

Consisting Essentially of: A minor variation of "consisting" is permissible, that is, an inconsequential amount of another element may be added.

Consisting of: A phrase limiting an invention to the specific elements recited. (Adding another element will usually avoid such a claim.)

Contributory Infringement: Selling a component of a patented assembly or a material for use in practicing a patented process that is especially adapted for the patented assembly or process, done with knowledge of the infringing use.

Continuation: *See* Application, Continuation.

Continuation-in-Part: *See* Application, Continuation-in-Part.

Convention, EPO or EPC: The European Patent Organization (EPO) or the European Patent Convention (EPC) is a group of 12 European countries wherein a single patent application is filed and examined as a unit. When the patent is allowed in the EPO, national patents can be filed, and are usually automatically allowed and issued in each designated European country.

Convention, International, Paris: A treaty between many of the nations of the world, the most important aspect being that an applicant in a Convention country may file in another Convention country within 1 year and receive the benefit of his filing date in the first country.

Convention, Patent Cooperation Treaty: The Patent Cooperation Treaty (PCT) convention operates as a "holding pattern" to preserve rights in designated foreign countries for a specified time. Chapter I will preserve rights for 20 months from the earliest priority filing. A timely filed Chapter II Demand will preserve patent rights in designated foreign countries up to 30 months from the earliest priority filing date.

Convention Date: The date of the application first filed in a Convention country provided it is not more than 1 year prior to filing in the country wherein such date is to be claimed.

Corroboration: Evidence from a person other than the inventor that supports and proves the inventive acts.

Count(s): Claim(s) made in common by two or more opposing parties in an interference, that is, the contested subject matter of the interference.

Court of Customs and Patent Appeals (C.C.P.A.): The historical court that reviewed U.S. Patent and Trademark Office decisions and customs cases, replaced by the Court of Appeals of the Federal Circuit (*which also see*).

Cross-Licenses: Licenses by and between separate owners of two or more patents whereby each receives a license under the patent or patents or certain patent claims of the other.

Declaration: An alternative procedure to sworn statements in patent application matters.

Dedication: Express or implied surrender to the public of an actual or potential patent right.

Defensive Publication: A publication and dedication to the public of a pending U.S. patent application.

Dependent Claim: A claim that refers back to and further restricts a preceding claim; includes the limitations of the prior claim.

Deposition: The testimony of a witness taken under oath before a notary public or other officer of a court, reduced to writing and authenticated, intended to be used upon the trial of an action.

Derwent: An important information source of foreign and U.S. patents.

Design Patent: A special kind of patent, directed to the ornamental characteristics of an object.

Diligence: That activity of an inventor following his conception of the invention and leading to its physical reduction to practice, or the filing of an application thereon, which in a contested case entitles recognition of his conception date as the date of invention for priority purposes.

Direct Infringement: Making, using, or selling a patented invention during the term of the patent.

Disclaimer: The renunciation of a patent claim. A disclaimer may be filed in the U.S. Patent and Trademark Office by a patentee at any time.

Disclosure: (1) A description of an invention; (2) what is described in a patent application.

Disclosure Document Program: A recent service of the U.S. Patent and Trademark Office to record information for 2 years within the Office to establish dates of conception, reduction to practice, progress, improvements, etc. Caution: No patent protection is established.

Discovery: In litigation or interference proceedings, the process of obtaining pertinent information from the opponent's records under court order.

District Court: Any of the 10 federal district courts within the United States where a lawsuit for legal relief in a patent matter can be filed.

Division: A second application derived from a patent application found by the patent examiner to include two or more independent and distinct inventions. The division will claim an invention that was not permitted to remain in the first application. *See also* Restriction.

Doctrine of Equivalents: An expansion of the literal language of claims so that one who has made inconsequential changes in the product or process to avoid infringement will nonetheless be an infringer.

Dominated: Covered by one or more claims of a patent.

Double Patenting: An improper attempt to obtain a second patent on the same invention, or an obvious variant as claimed in a first issued or pending patents. *See* Terminal Disclaimer.

Drawing : One or more specially prepared figures filed as part of a patent application to explain and describe the invention.

Duty of Disclosure: The requirement imposed by all persons involved with the patenting process to disclose information (patents, articles, laboratory data, etc.) to the patent examiner that may be material to the granting of a U.S. patent. *See* Fraud on the Patent Office, Inequitable Conduct.

Effective Date: Date as of which a reference is or would be applied against an application; date as of which an application or patent operates as such. A U.S. patent is effective as of its filing date as a reference against another U.S. application, and as a prior patent on its issue date. *See also* Convention Date.

Election: Choice by applicant of a claim or claims to be prosecuted, as made among or between divisions or groups set up by the patent examiner in the absence of a generic claim or of an allowable generic claim. *See* Restriction.

EPO: European Patent Organization. *See* Convention, EPO or EPC.

EPC: European Patent Convention. *See* Convention, EPO or EPC.

European Patent Convention or Organization: *See* Convention, EPO or EPC.

Examiner: An official of the U.S. Patent and Trademark Office charged with determining the patentability of applications.

Example: A detailed description of one, and only one, embodiment of an invention; often more precisely referred to as a "specific" or "working" example in contradistinction to alternatives referred to "by example" without detail.

Execute: To complete; with reference to an application, to sign an oath before a notary public or sign a declaration, thereby swearing to the averments of the oath required by law.

Ex Parte: Nonadversary legal proceedings (for one party only) as in a normal patent prosecution; contrasted to interference or opposition proceedings where two or more parties are opposed.

Experimental Use: Use primarily for purposes of testing the invention and not for profit or commerce.

Extension of Time: Permission by the U.S. Patent and Trademark Office to file a response to an official action late when accompanied by a timely filed petition and a fee.

Fee, Filing: A fee required by the U.S. Patent and Trademark Office at the time of filing of the patent application, or shortly thereafter with the additional payment of a surcharge.

Fee, Issue: Payment to the U.S. Patent and Trademark Office required by law for issuance of an allowed application; must be tendered within 3 months of allowance.

Fee, Maintenance: A fee payable to the U.S. Patent and Trademark Office to keep a patent enforceable. The U.S. fee is now payable at $3 \frac{1}{2}$, $7 \frac{1}{2}$, and $11 \frac{1}{2}$ years from date of issue. A late fee (only within 6 months) is permitted with the additional payment of a surcharge.

Fee, Petition: A fee required by the U.S. Patent and Trademark Office for certain actions.

File History (File Wrapper): The complete file of a patent application and all related documents prepared by the U.S. Patent and Trademark Office or submitted by the applicant during the prosecution of the application.

File Wrapper Estoppel: An admission by an applicant during the prosecution of the patent application. Its effect is that certain claims are not patentable or that certain claims are subject to a specific and restricted interpretation. This is also referred to as prosecution history estoppel.

Filing Date: The date when the application reaches the U.S. Patent and Trademark Office in complete form. An application can be filed without signatures of one or all of the inventors to obtain a filing date and a serial number. However, the properly executed declaration(s) need to be filed within 2 months or later with the simultaneous filing of a petition for extension of time and fee.

First to File: The present patent law in all foreign countries. The applicant who is the first to file will be awarded the patent over all others.

First to Invent: The present patent law followed only in the United States, which will award a patent to the first to invent. *See* Interference.

Forfeited Application: An allowed application on which the issue fee or maintenance fee has not been paid within the required period.

Fraud on the Patent Office: A breach of the duty of disclosure and/or candor by any one of the parties involved with the patent process. If found, the entire patent in question is held unenforceable, not specific claims. Subsequently issued divisional or continuation applications may also be "tainted".

Harmonization: The coming harmonizing alignment of all the patent laws of the countries of the world to produce substantially one world patent law system.

IIB: The Institut International des Brevets, a patent search and examination office in The Hague.

Indemnity from Suit: An agreement between the parties of interest that the patent holder will not sue regarding potentially infringing activity.

Inducement to Infringe: An act that induces another to infringe. For example, a recommendation made to a customer to practice a patented method.

Inequitable Conduct: Improper action by a patentee before the U.S. Patent and Trademark Office. If a breach of the duty of disclosure is found, the entire patent in question is held to be unenforceable.

Infringe: To make, use, or sell the patented item or process within the country covered by the patent, for example, the United States. Because the claims measure the invention, the object must contain at least every element of the claim before it infringes the invention.

INPADOC: The International Patent Documentation Center, Vienna, Austria. A source of information about patents of more than 25 countries.

Interference: The U.S. Patent and Trademark Office proceeding that determines the priority of invention between two or more applications or patents claiming the same invention.

Intervening Rights: A defense to infringement. When a patent is later reissued with broader claims because of inadvertent errors in claiming in the original patent, any person who practiced the broadened claims prior to the reissue can continue. For such a person, infringement would result only from practicing the surviving original claims.

Interview: A conference with the U.S. Patent and Trademark Office examiner either in person or by phone.

Invalid: Ineffective or void in law; a patent that is void for one or more statutory reasons, for example, lack of novelty, lack of invention, or inoperativeness.

Invention, Abandoned: An unexploited invention on which no patent application is filed for a long, unexplained time, usually during which others have entered the field.

Inventive Entity: The inventor or inventors.

Inventor, First: The person who has the earliest date for an invention that he or she has not explicitly or implicitly abandoned; he or she who first conceives and first reduces to practice or, if first to conceive but last to reduce to practice, was diligent prior to the entry of others into the field.

Inventor, Joint: One who, in conjunction with one or more others, conceives of an invention by an original, contributing mental process and who plays a part in causing the same to be reduced to practice.

Issue Fee: *See* Fee, Issue.

Kokai: An unexamined Japanese patent application. The first two digits of the application number indicate the year in the current emperor's reign. The first two digits plus 25 give the year of publication in the Showa (Hirohito) era (1925 to early 1989), for example, J59—— has a 1984 publication date. The first two digits plus 88 give the year of publication of more recent patent applications, for example, J01—— has a 1989 publication date.

Kokoku: An examined and allowed Japanese patent application. The first two digits indicate the Gregorian calender year of publication.

Label License: When a composition or product is sold by the owner or a licensee of a patent on the process of using the item sold, he or she may include on the label notice to the purchaser that the price includes a license to practice the process.

License: Any conveyance of a right under a patent that does not amount to an assignment. A license gives the licensee no title in the patent. Can be exclusive or nonexclusive.

Machine: One of the statutory types of inventions; ordinarily employed in the usual sense, that is, to designate a combination of mechanical elements.

Maintenance Fee: A fee required by the United States and most foreign countries to maintain a patent in force. It may be annual or periodic.

Manufacture: A term used in the patent statutes to cover any material thing in between an unshaped composition of matter and a machine. Examples are a bottle cap, incandescent bulb, and coated fabric.

Marking, Patent: The incorporation of one or more U.S. patent numbers to the surface of a patented article or to the surrounding package or packaging.

Markush Group: A form of claim reciting an element "Selected from the group consisting of . . . "; this is in effect a limited form of a generic claim. *See* Claim, Markush.

Mesne Assignment: A transfer of patent rights through one or more intermediate assignees rather than directly from the inventor.

New Matter: Matter attempted to be added to an application by amendment but that is refused on grounds of not being present, explicitly or implicitly, in the original disclosure.

Nonassertion Letter: A written communication by the owner of patent rights stating that he or she does not intend at that time to enforce his or her legal right to exclude others from practicing specified claims of one or more patents.

Notice of Allowability: An action by a U.S. patent examiner that the patent application has been placed in condition for allowance, and specific claims are allowed.

Notice of Allowance and Issue Fee Due: An action by a patent examiner that the claims (as presently amended) are allowed and the issue fee must be timely submitted within 3 months.

Notice of Appeal: *See* Appeal, Notice of.

Novelty: A requirement for patentability; the invention in its entirety must not have been known or used by others before its invention by the inventor.

Oath, Inventor's: Oath made by applicant as part of the application and averring that he or she believes himself or herself to be the first and original inventor of the subject matter for which a patent is sought. *See* Declaration.

Octrooi: A patent (Dutch or Flemish).

Offenlegungschrift (OLS): A published (unexamined) German patent application.

Office Action or Official Action: Ordinarily, a communication from the U.S. Patent and Trademark Office to the applicant or counsel stating the conclusions of the patent examiner as to patentability of the claims presented.

O.G.: *Official Gazette* of the U.S. Patent and Trademark Office. Published weekly.

One-Year Grace Period: Presently, a statutory rule of the U.S. Patent and Trademark Office to the effect that a valid patent can be granted on an application filed up to 1 year after public use or sale in the United States or patenting or publication anywhere in the world, provided the invention of the application can be shown to have been made before the public use, sale, patenting, or publication by another.

Opinion, Infringement: An opinion by an attorney as to the probable holding of a court on whether or not a specified product or process or apparatus would infringe one or more claims of a patent.

Opinion, Patentability: An opinion by an agent or attorney as to the probable holding by the U.S. Patent and Trademark Office or by a court as to allowability of claims to specified subject matter in view of the prior art, prior sale, offer for sale, etc.

Opinion, Validity: An opinion by an agent or attorney on whether or not a court would be likely to uphold a patent or patent claim in view of identified prior art or other specified facts indicating a lack of novelty, invention, originality, etc.

Opposition: A procedure available in many foreign countries by which third parties may formally oppose the grant of a patent.

Organisms, Living: Now proper U.S. patentable subject matter under 35 U.S.C. 101.

Organisms, Nonhuman Mammal: Now proper U.S. patentable subject matter under 35 U.S.C. 101.

Patent Agent: An individual having scientific or engineering background or experience registered to practice before the U.S. Patent and Trademark Office. The registration is obtained by passing a 1-day written examination administered two times a year by the U.S. Patent and Trademark Office at various locations about the country. The patent agent is subject to the same American Bar Association rules of ethics as is an attorney.

Patent Attorney: An individual admitted and licensed to practice before the bar of at least one state or the District of Columbia. The patent attorney does not necessarily have scientific background, although it helps. The patent attorney is usually, but not necessarily, admitted to practice before the U.S. Patent and Trademark Office.

Patent Cooperation Treaty (PCT): *See* Convention, Patent Cooperation Treaty.

Patent Marking: *See* Marking: Patent.

Patent Pending: A marking applied to an article of commerce to indicate that application has been made for a patent thereon; the situation in which an application has been filed and is under prosecution. Prohibited in many foreign countries.

Patent Search: A search of the prior art, for example, the open literature and U.S. and foreign patents.

Patentschrift: A Patent (German).

PCT: Patent Cooperation Treaty. *See* Convention, Patent Cooperation Treaty.

Person Skilled in the Art: The fictitious person supposed to have ordinary information and skill in the particular field to which a given invention pertains.

Petition: (1) Portion of an application praying for grant of a patent; (2) a request to the Commissioner of Patents that he or she exercise supervisory authority on a procedural or jurisdictional matter in the U.S. Patent and Trademark Office.

Petition for Extension of Time: For certain official actions within the U.S. Patent and Trademark Office, the time for response may be extended by the timely filing of a petition for extension of time and the appropriate fee.

Petition to Revive: When a U.S. patent application is inadvertently abandoned, under certain circumstances, a timely filed petition to revive may be successful to revive the patent application.

Plant Patent: An enforceable U.S. patent for a new asexually produced plant species or variety.

Power of Attorney: Grant of authority to a patent attorney or patent agent to transact all U.S. Patent and Trademark Office business connected with a designated patent application; usually combined with the formal petition for grant of a patent.

Preliminary Amendment: An amendment to a pending patent application prior to a patent examiner's action on the merits, or an amendment filed simultaneously with the filing of a patent application.

Preliminary Statement: A statement under oath in an interference proceeding setting out acts material to the origin with the inventor of the subject matter of the interference counts and the dates or approximate dates of such acts.

Prior Art: All prior knowledge relating to the claimed invention, including prior use, publications, and patent disclosures.

Prior Art Statement: A timely submission (usually within 3 months of the filing date) by the patent agent, patent attorney, or inventor to the patent examiner of the articles, references, patents, and laboratory results considered to be material to the determination of patentability of the subject matter of the patent application.

Priority Document: A certified copy of an application filed in another country, submitted in order to claim the benefit of the filing date under the International Convention.

Process: (1) One of the statutory types of invention; (2) the steps involved in changing the physical or chemical characteristics of a material. Such steps, as well as uses of new or known materials, are patentable if unobvious.

Proprietary Information: Information that is not generally known and of value competitively, for example, trade secrets and business information not in general circulation.

Prosecution: The progression of steps in the examination of the pending patent by submission of the prior art statement, the first official action, amendment and response to the first official action, review of the response by the patent examiner, issuance of a final rejection, response after final rejection, appeal notice, notice of allowance, etc.

PTO: The U.S. Patent and Trademark Office.

Publication: Any document, including patents of most countries, that is printed in the ordinary sense (published) and is actually or presumptively available to the public; includes other documents, not necessarily printed, that are readily available to workers in a given art, for example, advance abstracts of journal articles, treatises deposited in public libraries, or government literature.

Public Sale: Sale or proffered sale (a brochure) of a given article or product (or process) to a member of the public at large. *See* Bar, Statutory.

Public Use: Utilization of an invention in a nonexperimental manner open to the public, that is, under conditions permitting observation by others not under obligation of secrecy. *See* Bar, Statutory.

Reads on: A term used to mean that a claim includes within its scope certain subject matter.

Reduction to Practice, Actual: Carrying out the inventive concept in a tangible way; the first successful experimental demonstration.

Reduction to Practice, Constructive: Written disclosure (paper, conceptual) of an invention in an application, as opposed to actual reduction to practice by experimental trial. Such application or a continuation thereof claiming the invention may issue.

Reexamination: A recent statutory post-issuance reexamination of a issued patent to provide the requester (the patent holder, a competitor, etc.) an opportunity to cite newly discovered prior art patents and/or printed publications to evaluate the patentability of the claims in the original patent over the newly discovered art.

Reference: Any publication disclosing subject matter that is or may be pertinent to consideration of the patentability of a specified process, product, or group of products. More specifically, a printed publication cited by the U.S. Patent and Trademark Office as bearing on patentability of claims presented.

Registered to Practice: An attorney or agent who has passed the U.S. Patent and Trademark examination, undergone and passed review as one morally qualified to practice, and has paid the fees necessary to be placed on the U.S. Patent and Trademark Office Register.

Reissue: A new patent issued by the U.S. Patent and Trademark Office to correct inadvertent errors in the claims of the original patent or to seek reconsideration in view of newly found prior art. The patent term is not changed. The claim language may be broadened (only within 2 years) or limited.

Rejection: Refusal to allow a patent claim; the U.S. Patent and Trademark Office action holding claims unpatentable.

Rejection, Final: A decision of the U.S. Patent and Trademark Office examiner finally rejecting one or more claims of an application. The response by the applicant must overcome all objections if it is to be accepted and written into the file. The response may be the basis for an appeal, or it may be the appeal itself.

Request for Reconsideration: Further argument submitted after a claim has been rejected.

Restriction: A requirement that a patent application containing more than one distinct invention be restricted to only one of the inventions disclosed. *See also* Division.

Serial Number: An identifying number given by the U.S. Patent and Trademark Office to all complete patent applications as of the day they are received or made complete.

SIR: *See* Statutory Invention Registration.

Specification: That part of a patent that describes the invention and how to make and use it.

Statutory Invention Registration (SIR): The Commissioner of Patents and Trademarks at the request of the patentee, can publish the SIR. This document is not examined, and is for the establishment of prior art over another applicant. The abstract is published in the *Official Gazette,* and the contents are available from the U.S. Patent and Trademark Office. The applicant irrevocably waives the right to receive a U.S. patent.

Statutory Period: The period within which a response must be made to an office action or an appeal taken therefrom, if a holding of abandonment is to be avoided; less commonly, the period within which an application must be filed.

Terminal Disclaimer: A document filed in one or more patents having a common owner (assignee) disclaiming certain claims or disclaiming a portion of the patent term. This is usually required to overcome a judicially created double patenting rejection.

Testimony: Evidence given by a witness, under oath or affirmation, as distinguished from evidence derived from writings and other sources.

Trademark: A U.S. grant to an individual or to an organization to establish the source of goods in commerce.

Trade Secret: A business practice kept proprietary to an individual or an organization for the purpose of obtaining or maintaining a competitive advantage in the marketplace.

Traverse: To dispute; to take issue with a holding of a patent examiner.

Unenforceable: This term is used when a U.S. patent, although obtained and issued from the U.S. Patent and Trademark Office, is determined by a court to be obtained improperly under sections of 35 U.S.C. The entire patent is declared to be unenforceable against any third infringing parties.

Unobvious: An invention that would not be obvious to one having ordinary skill in the subject area at the time the invention was made, assuming knowledge of all prior art.

U.S.C.: United States Code. Title 35 includes rules and regulations for patents, trademarks, and copyrights.

Useful: Serving a purpose, especially a valuable, as opposed to a frivolous, fraudulent, or immoral one.

Utility: One of the three general statutory requirements of invention; fitness for some desirable practical purpose.

Valid: Sound and justified; meeting the tests of patentability.

Venue: County or district in which a legal action, for example, the making of an oath, the filing of a suit, or the execution of a patent application, is to be performed.

WIPO: The World Intellectual Property Organization, Geneva, a part of the United Nations. A coordinating body for patent systems and procedures.

Worked; Working: Numerous foreign countries require "working" of a patent within a specified time after grant. Conditions for satisfying "working" requirements vary widely, from publication of availability of licenses to the patent to actual manufacture or sale of a product claimed in the invention.

Bibliography

The U.S. Patent System

Patent Laws. Superintendent of Documents, U.S. Government Printing Office, Washington, DC 20402. *The complete U.S. patent law, indexed.*

General Information Concerning Patents. Superintendent of Documents, U.S. Government Printing Office, Washington, DC 20402. $4.00. Stock Number 003-004-00522-0. *A brief but thorough summary of the U.S. patent system, requirements for obtaining a patent, and services provided by the PTO. There is a comparable bulletin on trademarks.*

Code of Federal Regulations Title 37. Patents, Trademarks, and Copyrights. Superintendent of Documents, U.S. Government Printing Office, Washington, DC 20402. $16.00. *A complete codification of the rules of practice for obtaining patents, trademarks, and copyrights, with related documents.*

Manual of Patent Examination Procedure, M.E. U.S. Department of Commerce, Patent & Trademark Office, Superintendent of Documents, P.O. Box 1533, Washington, DC 20013. *A loose-leaf manual covering all aspects and details of the examination of patents. Continually revised.*

Attorneys and Agents Registered in Practice Before the U.S. Patent and Trademark Office. U.S. Department of Commerce, Patent and Trademark Office, $10.00, for sale by the Superintendent of Documents, U.S. Government Printing Office, Washington, DC 20402. *The title describes it.*

Patent Law

Patent Law Fundamentals, 2nd ed., by Peter D. Rosenberg, Clark Boardman Co.: New York, 1984. *A thorough but concise and readable book that*

1997–4/91/0163/$06.00/1
© 1991 American Chemical Society

discusses and interprets all aspects of patenting, patent law, and exploitation of patent rights.

Patents, by Donald Chisum, Matthew Bender & Co.: New York, 1989 (annually updated). *A comprehensive multivolume treatise, often quoted by the courts, covering details of patent practice and litigation.*

Patents, Trademarks, and Related Rights, by Stephen P. Ladas, Harvard University Press: Cambridge, MA, 1975. *A useful historical three-volume, encyclopedic work on all aspects of patent philosophy and practice, worldwide.*

The Encyclopedia of Patent Practice and Invention Management, Robert Calvert, Ed., Reinhold: New York, 1964. *Authoritative articles on patent matters and management. Very useful for background and history.*

The Duty of Candor Under Rule 56 and the Evolution of Proposed Rule 57. American Intellectual Property Law Association (AIPLA) Quarterly Journal, Volume 16, Number 1, 1988. AIPLA, Suite 203, 2001 Jefferson Davis Highway, Arlington, VA 22202. *An informative and sobering article about the high duty of candor in patent matters.*

Candor in Prosecution: A Monograph on Fraud and Candor in Prosecution Before the Patent and Trademark Office, Margaret Boulware and Frank Robbins, Eds., American Intellectual Property Law Association (AIPLA) Chemical Practice Committee, 1985. AIPLA, Suite 203, 2001 Jefferson Davis Highway, Arlington, VA 22202. *Exactly as the title describes it.*

Patent Law Perspectives, 2nd ed., Vol. 1, by Donald R. Dunner et al., Matthew Bender & Co.: New York, 1989. *Accounts of a former Commissioner of the U.S. Patent Office.*

Patents, A Treatise on the Law of Patentability, Validity, and Infringement, by Donald S. Chisum. Matthew Bender & Co.: New York, August 1989. *A very useful reference for patent agents and attorneys.*

Patent Law for the Nonlawyer: A Guide for the Engineer, Technologist, and Manager, by Burton A. Amernick, Van Nostrand Reinhold: New York, 1986. *The title describes it.*

Procedures in Obtaining Patents

Introduction to Patents (an audio course), by Richard Racine, Esq., American Chemical Society: Washington, DC, 1989. *A 6-hour audio cassette course on patent matters for the chemist; with study guide.*

Patent It Yourself, 2nd ed., by David Pressman, Nolo Press (950 Parker Street): Berkeley, CA, 1988. *A text of general interest to the chemist.*

Patent Information

World Patents Index, World Patents Abstracts, Chemical Patents Index, Derwent/SDC On-Line Search Service, Derwent Publications Ltd., 128 Theobalds Road, London, England WC1X 8RP. *Patent abstract and search services. Catalogues, indexes, and instruction manuals are issued annually.*

Basic Chemical (Patent) Practice. American Intellectual Property Law Association (AIPLA) Chemical Practice Committee, 1989. AIPLA, Suite 203, 2001 Jefferson Davis Highway, Arlington, VA 22202. *An annual summary and seminar of recent important aspects of chemical patents..*

Basic Chemical and Biotechnology Practice. American Intellectual Property Law Association (AIPLA) Chemical Practice Committee, 1988. AIPLA, Suite 203, 2001 Jefferson Davis Highway, Arlington, VA 22202. *A useful discussion of aspects of biotechnological patents.*

Patent Trademark and Copyright Journal, Bureau of National Affairs, 1231 25th Street, N.W., Washington, DC 20037. *A weekly update of recent cases in patent, trademark, and copyright law.*

Biotechnology Patent-Related Matters

Patents in Chemistry and Biotechnology, by Phillip W. Grubb, Oxford University Press, Walton Street, Oxford, England OX2 6DP, 1986. *An excellent 335-page basic book describing generally chemical and biological patenting in the United States and the United Kingdom.*

Biotechnology and Patent Protection: An International Review, by F. K. Beier, R. S. Crespi, and J. Straus, Organization for Economic Cooperation, Paris, France, 1985. *An overview of worldwide biotechnological patent practices and requirements.*

Patenting in the Biological Sciences: A Practical Guide for Research Scientists in Biotechnology and the Pharmaceutical and Agrochemical Industries, by R.S. Crespi, Wiley: New York, 1982. *The title describes it.*

Patent and Trademark Tactics and Practice, 2nd. ed., by David A. Burge, Wiley: New York, 1984. *The title describes it.*

Protecting Biotechnology Inventions: A Guide for Scientists, by Roman Saliwanchik, Science Tech Publishers, 701 Ridge Street, Madison, WI 53705, 1988. *A useful text to provide an overview for chemists.*

New Developments in Biotechnology: Patenting Life: Special Report, U.S. Congress, Office of Technology Assessment, OTA-BA-370, U.S. Government Printing Office, Washington, DC 20402, April 1989. *A government compilation of U.S. biotechnological patents*

The Value of Patents

The Employed Inventor in the United States, by Fredrik Neumeyer, MIT Press: Cambridge, MA, 1971. *A now dated analysis and discussion of a study of compensation practices for inventors employed by major industries, government, and universities.*

The Sources of Invention, 2nd ed., by John Jewkes, David Sawers, and Richard Sillerman, Norton: New York, 1969. *A discussion of the factors affecting the creativity of individual and employed inventors, with case histories of a number of important inventions.*

Invention and Economic Growth, by Jacob Schmookler, Harvard University Press: Cambridge, MA, 1966. *A report of a 20-year study of invention and market development activity as related to patent statistics.*

Business Aspects of Patents

McCarthy's Desk Encyclopedia of Intellectual Property, J. Thomas McCarthy, BNA Books Distribution Center, 300 Raritan Center Parkway, P.O. Box 7816, Edison, NJ 08818–7816, 1991. *A very useful volume for any chemist or patent practitioner.*

Trade in Innovation, by J. A. D. Cropp, D. C. Harris, and E. S. Stern, Wiley-Interscience: London, 1970. *A concise discussion of the value of patent protection and of the basic principles of exploiting patent rights.*

Corporate Caution and Unsolicited New Project Ideas, by Del I. Hawkins and Gerald G. Udell, *J. Pat. Off. Soc.* **1976,** *58,* 375. *A survey of corporate waiver requirements relative to unsolicited suggestions, with comment.*

Employee Agreements *and* **Trade Secrets . . . Ethics and Law.** American Chemical Society, 1988. *Informative pamphlet prepared by the ACS Committee* *on Professional Relations that summarizes the basic principles of employee–employer relationships with respect to patent ownership and proprietary information.*

Foreign Patent Systems and Procedures

PCT Applicant's Guide, PCT Receiving Office, Box PCT, Washington, DC 20231. *The authoritative loose-leaf set of PCT requirements.*

Manual for the Handling of Applications for Patents, Designs, and Trademarks throughout the World, Octrooibureau Los en Stigter, Amsterdam. *A comprehensive summary of the law and procedures of every country that has a patent and/or trademark system. Loose-leaf, regularly updated.*

Convention on the Grant of European Patents, published by the Government of the Federal Republic of Germany, 1973. *Complete text of the documents signed at Munich setting up a European System for the Grant of Patents. With associated documents relating to establishment of the European Patent Office. (Historical value)*

Copyright

Nimmer on Copyright, (updated annually), Melville B. Nimmer, Ed., Matthew Bender & Co.: New York. *A loose-leaf treatise often cited by the courts.*

Copyright Law Journal (published monthly), Niel Boorstyn of McCutchen, Doyle, Brown, & Enersen, 3 Embarcadero Center, San Francisco, CA 94100. *A useful service for the attorney, manager, or scientist involved in copyright matters*

Copyright: Principles, Law, and Practice, Vols. I, II, and III, Paul Goldstein, Ed., Little, Brown, and Company, Law Division, 34 Beacon Street, Boston, MA 02108, 1991.

Trade Secrets

Milgrim on Trade Secrets, (loose-leaf, updated periodically), Roger M. Milgrim, Matthew Bender & Co.: New York, 1989. *A useful service often cited by the courts.*

Administrative Law Review: Your Business, Your Trade Secrets, and Your Government, Spring 1982, Volume 34, Number 2. *A seminar on protecting and obtaining commercial information from the government.*

Patent Anecdotes and Innovation History

National Invention Center, National Inventors Hall of Fame, 80 W. Bowery, Suite 201, Akron, OH 44309, (216) 762—1990. *A variety of Services including information about inventions and U.S. inventors inducted into the Hall.*

Patent Pending, by Richard L. Gausewitz, Alson Publishing Co.: Long Beach, CA, 1983. *Useful (and not so useful) historical patents.*

American Dreams: 100 Years of Business Ideas and Innovation from the Wall Street Journal, by K. Morris et al., Lightbulb Press, Inc., 1185 Avenue of the Americas, New York, NY 10036, 1990. *One to two page depictions of U.S. patents and innovations.*

Extraordinary Origins of Everyday Things, by Charles Panati, Harper and Row, 10 East 53rd Street, New York, NY 10022, 1987. *An interesting general volume describing the creation of many U.S. articles of commerce.*

Inventors at Work, Interviews with 16 Notable American Inventors, by Kenneth A. Brown, Tempus Books of Microsoft Press, a Division of Microsoft Corporation, 16011 NE 36th Way, Box 97017, Redmond, WA 98073–9717, 1988. *The title describes it.*

Organizations Mentioned

American Chemical Society

1155 16th Street, N.W.
Washington, DC 20036
(202) 872–4600
(800) ACS–5558
Fax (202) 872–6067
Telex 440159

American Petroleum Institute (API)

Central Abstracting and Indexing Service
275 7th Avenue
New York, NY 10001
(212) 366–4040
Fax (212) 366–4298

Chemical Abstracts Service

2540 Olentangy River Road
Columbus, OH 43202–1505
(614) 421–6940
(800) 848–6538
Fax (614) 447–3713
Telex: 6 842 086 CHMAB

Derwent Inc.

Suite 401, 1313 Dolley Madison Blvd.
McLean, VA 22101
(703) 790–0400
(800) 451–3451
Fax (703) 790–1426

1997–4/91/0169/$06.00/1
© 1991 American Chemical Society

Derwent Publications, Ltd.

Rochdale House
128 Theobalds Road
London, England WCIX 8RP
Fax 011–4471 405 3630

Dialog Information Systems, Inc.

3460 Hillview Avenue
Palo Alto, CA 94304
(415) 858–3785
(800) 3DIALOG, (800) 334–2564
Fax (415) 858–7069

European Patent Office

DG1, P.O. Box 5818
2280 HV Risjswijk
The Netherlands
(070) 3402040
Fax (070) 3403016
Telex 31651 EPO NL

European Patent Office, Vienna Sub-Office

Mollwaldplatz 4, A-1040
Vienna, Austria
(+431) 50155
Fax (+431) 5053386
Telex 136337 impa a
(U.S. Representative: IFI/Plenum Data Company, which see)

IFI/Plenum Data Company

302 Swann Avenue
Alexandria, VA 22301
(703) 683–1085
Fax (703) 683–0246

Industrial Opportunities, Ltd.

Homewell, Havent, Hampshire
England PO9 1EF

Institut International des Brevets (IIB)

Patentlaan 2
Rijswijk (ZH),
The Netherlands

Institute for Scientific Information

3501 Market Street
Philadelphia, PA 19104
(215) 386–0100
Fax (215) 386–6362
Telex 84–5305

International Patent Documentation Center (INPADOC)

Mollwaldplatz 4, A-1040
Vienna, Austria
U.S. Representative: IFI/Plenum Data Company
(which see) Attn.: James Menge, Account Executive

Licensing Executives Society (U.S.A.) Inc.

71 East Avenue,
Suite S
Norwalk, CT 06851
(203) 852–7168
Fax (203) 838–5714
    For Publications: LES Publications
    c/o Studeny Co.
    544 William Penn Hotel
    Pittsburgh, PA 15219

Patent Resource Group, Inc.

2000 Pennsylvania Avenue, N.W.
Washington, DC 20006
(202) 223–1175
Fax (202) 872–0890

Practicing Law Institute

810 7th Avenue
New York, NY 10019
(212) 765–5700
Fax (212) 265–4742

Research Corporation (Grants and Information)

6840 E. Broadway Boulevard
Tucson, AZ 85710–2815
(602) 296–6771
Fax (602) 721–8789

Research Corporation and Technology
(Technology Transfer)

6840 E. Broadway Boulevard
Tucson, AZ 8571–2815
(602) 296–6400
Fax (602) 296–8157

University Patents, Inc.

1465 Post Road East
Westport, CT 06880
(203) 255–6044
Fax (203) 254–1102

Index

Copy editing and indexing: Janet S. Dodd
Production: Margaret J. Brown and Donna Lucas
Cover design and illustration: Neal Clodfelter
Acquisition: Robin Giroux
Printed and bound by Maple Press, York, PA

Highlights from ACS Books

Good Laboratory Practices: An Agrochemical Perspective
Edited by Willa Y. Garner and Maureen S. Barge
ACS Symposium Series No. 369; 168 pp; clothbound, ISBN 0–8412–1480–8

Silent Spring Revisited
Edited by Gino J. Marco, Robert M. Hollingworth, and William Durham
214 pp; clothbound, ISBN 0–8412–0980–4; paperback, ISBN 0–8412–0981–2

Insecticides of Plant Origin
Edited by J. T. Arnason, B. J. R. Philogène, and Peter Morand
ACS Symposium Series No. 387; 214 pp; clothbound, ISBN 0–8412–1569–3

Chemistry and Crime: From Sherlock Holmes to Today's Courtroom
Edited by Samuel M. Gerber
135 pp; clothbound, ISBN 0–8412–0784–4; paperback, ISBN 0–8412–0785–2

Handbook of Chemical Property Estimation Methods
By Warren J. Lyman, William F. Reehl, and David H. Rosenblatt
960 pp; clothbound, ISBN 0–8412–1761–0

The Beilstein Online Database: Implementation, Content, and Retrieval
Edited by Stephen R. Heller
ACS Symposium Series No. 436; 168 pp; clothbound, ISBN 0–8412–1862–5

Materials for Nonlinear Optics: Chemical Perspectives
Edited by Seth R. Marder, John E. Sohn, and Galen D. Stucky
ACS Symposium Series No. 455; 750 pp; clothbound; ISBN 0–8412–1939–7

Polymer Characterization:
Physical Property, Spectroscopic, and Chromatographic Methods
Edited by Clara D. Craver and Theodore Provder
Advances in Chemistry No. 227; 512 pp; clothbound, ISBN 0–8412–1651–7

From Caveman to Chemist: Circumstances and Achievements
By Hugh W. Salzberg
300 pp; clothbound, ISBN 0–8412–1786–6; paperback, ISBN 0–8412–1787–4

The Green Flame: Surviving Government Secrecy
By Andrew Dequasie
300 pp; clothbound, ISBN 0–8412–1857–9

For further information and a free catalog of ACS books, contact:
American Chemical Society
Distribution Office, Department 225
1155 16th Street, NW, Washington, DC 20036
Telephone 800–227–5558

32.00